T0336866

Elements of Logical Reasoning

Some of our earliest experiences of the conclusive force of an argument come from school mathematics: faced with a mathematical proof, we cannot deny the conclusion once the premisses have been accepted. Behind such arguments lies a more general pattern of 'demonstrative arguments' that is studied in the science of logic. Logical reasoning is applied at all levels, from everyday life to advanced sciences, and a remarkable level of complexity is achieved in everyday logical reasoning, even if the principles behind it remain intuitive. Jan von Plato provides an accessible but rigorous introduction to an important aspect of contemporary logic: its deductive machinery. He shows that when the forms of logical reasoning are analysed, it turns out that a limited set of first principles can represent any logical argument. His book will be valuable for students of logic, mathematics, and computer science.

JAN VON PLATO is Professor of Philosophy at the University of Helsinki. He is the author of *Creating Modern Probability* (Cambridge, 1994) and, with Sara Negri, *Structural Proof Theory* (Cambridge, 2001) and *Proof Analysis* (Cambridge, 2011).

Elements of Logical Reasoning

JAN VON PLATO

CAMBRIDGE
UNIVERSITY PRESS

University Printing House, Cambridge CB2 8BS, United Kingdom

One Liberty Plaza, 20th Floor, New York, NY 10006, USA

477 Williamstown Road, Port Melbourne, VIC 3207, Australia

314-321, 3rd Floor, Plot 3, Splendor Forum, Jasola District Centre, New Delhi - 110025, India

79 Anson Road, #06-04/06, Singapore 079906

Cambridge University Press is part of the University of Cambridge.

It furthers the University's mission by disseminating knowledge in the pursuit of education, learning and research at the highest international levels of excellence.

www.cambridge.org
Information on this title: www.cambridge.org/9781107036598

First published 2013

A catalogue record for this publication is available from the British Library

Library of Congress Cataloging in Publication data
Von Plato, Jan, author.
Elements of logical reasoning / Jan von Plato.
pages cm
Includes bibliographical references and index.
ISBN 978-1-107-03659-8 (hardback) – ISBN 978-1-107-61077-4 (paperback)
1. Logic, Symbolic and mathematical. 2. Reasoning. I. Title.
QA9.V66 2013
511.3 – dc23 2013039013

ISBN 978-1-107-03659-8 Hardback
ISBN 978-1-107-61077-4 Paperback

Contents

Preface

When I was little and Christmas time was approaching, we children knew that there would be two kinds of presents: the soft packages that contained useful but unexciting clothes, and the hard boxes that contained gorgeous new toys. I learned later that the same formula repeats itself often in life, and even in logic. There are the discussions about first principles: what rests on what, what comes first in the end of all analyses, and what it all means – and these are the useful but relatively unexciting soft packages. Then there is the box that is really interesting to open, and that is what I call the deductive machinery of logic – how it all actually works. Others have called it the inferential engine. I believe that logic should not be presented to us just in those soft packages – the hard box has to be there to be opened as well, so that we can find out how logical arguments function. It is a hands-on kind of learning in which one tries and retries things by oneself until the machinery runs smoothly. Then it is the time to discuss the nature of the first principles.

The book begins with a linear form of proofs that I learned from Dag Prawitz' Swedish compendium *ABC i Symbolisk Logik*. Little did I think, back in 1973 when using that text for the first time, that my teaching of elementary logic would one day grow into a comprehensive presentation in the form of a book. Over the years that I have taught logic, students too numerous to be listed here have added to my understanding of how the presentation of the topics should be structured. Next to these experiences, Sara Negri is the person who has contributed decisively to the direction of my work in logic in general. I want to dedicate this book to her and to our proof-theoretical adventure that began quite casually in 1997.

Cambridge University Press sent me, through the good offices of Hilary Gaskin, four anonymous referees' reports with overwhelmingly positive views and comments on the manuscript, which also led to numerous improvements in the presentation.

PART I

First steps in logical reasoning

1 Starting points

Some of our earliest experiences of the conclusive force of an argument come from school mathematics: Faced with a mathematical proof, however we try to twist the matter, there is no possibility of denying the conclusion once the premisses have been accepted.

Behind the examples from mathematics, there is a more general pattern of 'demonstrative arguments' that is studied in the science of logic. Logical reasoning is applied at all levels, from everyday life to the most advanced sciences. As an example of the former, assume that under some specific conditions, call them A, something, call it B, necessarily follows. Assume further that the conditions A are fulfilled. To deny B under these circumstances would lead to a contradiction, so that either B has to be accepted or at least one of the assumptions revised – or at least that is what the fittest thinker would do to survive.

A remarkable level of complexity is achieved in everyday logical reasoning, even if the principles behind it remain intuitive. We begin our analysis of logical reasoning by the observation that the forms of such reasoning are connected to the forms of linguistic expression used and that these forms have to be made specific and precise in each situation. When this is done, it turns out that a rather limited set of first principles is sufficient for the representation of any logical argument. What appears intuitively as an unlimited horizon of ever more complicated arguments, can be mastered fully by learning these first principles as explained in this book.

1.1 Origins

The idea of logical reasoning appears in the ancient 'science of demonstrative arguments', a terminology from the first logic book ever, Aristotle's *Prior Analytics.* Demonstrative arguments move from what is assumed to be **given** to a **thing sought**. The given can consist of a list of assumptions, the sought of a claim to be proved. A demonstrative science is organized as follows:

1. There are, first, certain **basic concepts** supposed to be understood immediately. Think, as an example, of points and lines in geometry, of a

point being incident with a line, and so on. Next there are **defined concepts**, ones that are not immediately understood. These have to be explained through the basic concepts. A triangle, say, can be defined to be a geometric object that consists of three straight line segments such that any two segments meet at exactly one common endpoint.

2. A second component of a demonstrative science consists of **axioms**. These are assertions the truth of which is immediately evident. We shall soon see some examples of ancient axioms. Next to the axioms, there are assertions the truth of which is not immediately evident. These are the **theorems** and their truth has to be reduced back to the truth of the axioms through a **proof**.

Proofs are things that start from some given **assumptions** and then proceed step by step towards a sought **conclusion**. The most central aspect of such a demonstrative argument is that the conclusion follows necessarily if the assumptions are correct. What the nature of this necessary following is, will be shown by some examples. We shall not, in general, aim at giving any exhaustive coverage of the various concepts that arise, but pass forward through examples. These are situations in which we have a good understanding of things.

1.2 Demonstrative arguments

Let us have a look at some examples of arguments in which the conclusion follows from some given assumptions.

(a) **Ancient geometry.** There are two types of situations in elementary geometry. In the first, we have some **given objects** of geometry such as points, lines, and triangles, with some prescribed properties. Next there is a **sought object** that has to have a prescribed relation to the given ones. Say, there are two given points, with the property that they are distinct, and the sought object is a triangle with the properties that the line segment with the given points as extremities is the base of the triangle, and that the triangle is equilateral. This is the first result to be established in Euclid's *Elements*, formulated as what is called a construction problem.

In a second kind of situation, there are also given objects with some properties, but the task is to simply prove that these objects have some additional new property. No explicit task of construction is mentioned, but the solution of the task to prove a property often requires intermediate steps of construction of auxiliary geometrical objects.

The following example comes from ancient Greek geometry and is of the second kind. Some of the terminology and notation is modern, but the geometrical argument in the example remains the same. Consider any given triangle with the three angles α, β, and γ. Then the sum of these angles is 180°. The result clearly is not anything the truth of which would be immediately evident, but a proof is required. The following figure illustrates the situation:

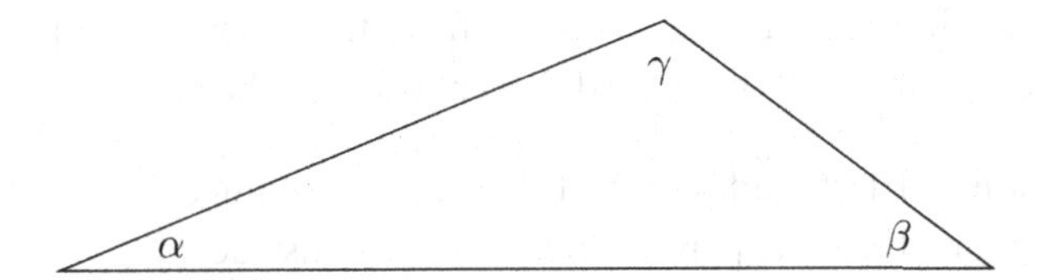

We have a base of the triangle, limited by the angles α and β. To prove the claim about the sum of the three angles, the sides are next prolonged and a **parallel** to the base drawn through the point that corresponds to angle γ. These are the auxiliary constructions needed:

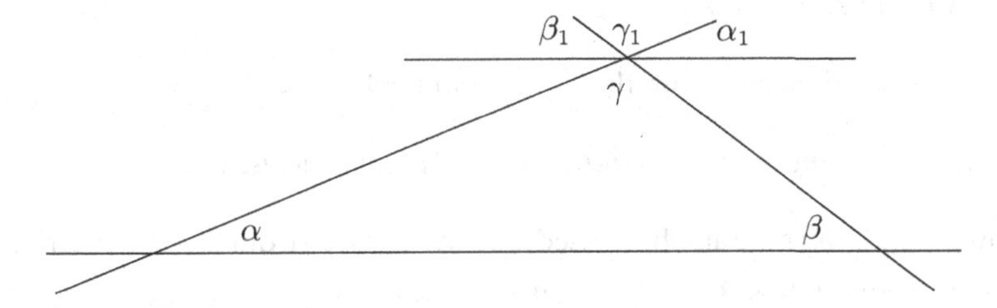

Symbols have been added to the figure, namely α_1, β_1, and γ_1. We reason as follows: The angle opposite to the original angle γ, namely γ_1, is equal to γ. Next, the line from the original angle α to angle γ intersects the base and the parallel to it. Therefore the angle that is marked by α_1 is equal to the lower left angle α of the triangle. Similarly, β_1 is equal to the lower right angle β of the triangle. We now see that α, β, and γ make up two right angles, or 180°.

The principles that were used in the proof were:

I. *The opposite angles of two intersecting lines are equal.*
II. *If a line intersects two parallel lines, the corresponding angles are equal.*

Both of these were taken to be immediate geometric truths in ancient geometry, i.e., they were considered axioms. If they are accepted, it seems that the claim about the sum of the angles of a triangle would not be a matter of opinion, but a **necessary consequence** of what has been assumed.

In addition to the axioms, what are called **construction postulates** were also used. These include the following, directly from the mentioned standard presentation of Greek geometry, Euclid's *Elements*:

III. *To continue a given finite straight line segment indefinitely.*
IV. *To draw a parallel to a given line through a point outside the line.*

We have a geometric configuration, some properties of which are assumed and with further properties that follow from the construction postulates. We go through in detail the steps that were taken in the proof:

1. By postulate III, the sides of the triangle are continued.
2. By postulate IV, a line parallel to the base is constructed.
3. By axiom II, $\alpha = \alpha_1$.
4. By axiom II, $\beta = \beta_1$.
5. By axiom I, $\gamma = \gamma_1$.
6. $\alpha_1 + \gamma_1 + \beta_1 = 180^\circ$.
7. By 3, 4, and 5, $\alpha_1 + \gamma_1 + \beta_1 = \alpha + \gamma + \beta$.
8. By 6 and 7, $\alpha + \gamma + \beta = 180^\circ$.

Step 8 is based on an axiom that is given in Euclid's *Elements* as:

V. *Any two things equal to a third are equal among themselves.*

Laws of addition have also been used, and in 6 it is seen from the construction that the three angles make up for two right angles.

(b) An example from arithmetic. One might think that perhaps the objects of geometry are too abstract and our intuitions about their immediately evident properties not absolutely certain. We can take instead the natural numbers: 0, 1, 2, ... Such a number is **prime** if it is greater than one and divisible by only one and itself: 2, 3, 5, 7, 11, 13 .... This series goes on to infinity. **Twin primes** are two consecutive odd numbers that are prime, say 5 and 7, 11 and 13, 17 and 19, and so on. Nobody knows if there is a greatest twin prime pair, or if their series goes on to infinity. Consider three consecutive odd numbers greater than 3. We claim that they cannot each be prime. Assume to the contrary this to be the case, i.e., assume there to be three numbers n, $n+2$, $n+4$ such that each is prime and $n > 3$. One out of any three consecutive numbers is divisible by 3 and, thus, one of n, $n+1$, $n+2$ is divisible by 3. By our assumption, it can be only $n+1$. But then also $n+1+3 = n+4$ is divisible by 3. Our assumption about three primes in succession turned out false.

There is a point in the argument that calls for some attention: It is essential to require that the three odd numbers be greater than 3. We concluded that for any n, one of n, $n + 1$, $n + 2$ is divisible by 3, and to further conclude that a number divisible by 3 is not prime, it needs to be distinct from 3. Indeed, the sequence 3, 5, 7 is excluded by the requirement.

There does not seem to be any place for opinions about the arithmetical truth established by the above argument. Someone might come and make a clever observation about a proof, especially if it was more complicated than the above example: Maybe something went wrong at some place in the proof. The thing to notice is that the very possibility of having made a mistake presupposes the possibility of the contrary, namely to have a correct proof.

(c) **An example from everyday life.** At Cap Breton in France, everyone agrees about the following rule: If the wind is hard, it is forbidden to swim. Here comes someone, in agreement with the rule, who also adds: I see some people swimming so I conclude that it is not forbidden to swim, even if I can see that the wind undoubtedly is hard. We could rebuke this someone: You accept that if the wind is hard, it is forbidden to swim. You also accept that the wind is hard. Therefore you accept that it is forbidden to swim, but you also deny it, which makes you contradictory. The person in question might say that it is not forbidden in any legal or moral way to hold contradictory opinions, nor is it a psychological impossibility. Whether it is disadvantageous in the struggle for survival can be debated.

Logical reasoning is based on the acceptance of certain criteria of rationality, such as not to both accept and deny a claim. Such accepting and denying may be hidden: If we accept a claim of a conditional form, say, if something A, then something B, but deny B, acceptance of A will lead to a contradiction as with the Cap Breton bather. The chain of inferences that leads to a contradiction can be so long that we do not necessarily notice anything. However, if a contradiction is pointed out, we should revise some of our assumptions.

1.3 Propositions and assertions

Logical reasoning operates on assumptions and what can be concluded from these. Assumptions and conclusions are things we obtain from **propositions.** We can also call them **sentences.** There is no use in trying to define what sentences are in general. We shall be content to have examples of

complete declarative sentences. Such a sentence expresses a **possible state of affairs**. Again, what possibility or state of affairs etc. is need not be explained in general, but we rest content with good examples. Consider the sentence *It is dark*. Whether this is correct depends on time and place, so let us assume they are fixed. Correctness may also depend on how one defines darkness, astronomically, in civil terms, or what have you, but that is not essential: We have paradigmatic examples of darkness and know what that means and we also know that the notion can be a bit hazy at times. Things such as the natural numbers and their properties would be less hazy, as in: One of 1733, 1735, and 1737 is divisible by 3, something we should believe in by the argument of Section 1.2.

A sentence is something neutral: It merely expresses a possible state of affairs. To make a sentence, call it A, into an assumption, we have to add something to A. This we do by stating: *Let us assume that* A. Similarly, if we conclude A, we actually make a **claim**, namely: A *is the case*. Thus, a sentence is turned into an **assertion** by the addition of an **assertive mood**: *It is the case that*.... Other such moods include the interrogative mood for making questions and the imperative mood for giving commands. The sentences that we utter come with a mood that is usually understood by the listener. We do not need to add in front of every sentence *it is the case that*..., even if we sometimes do it for emphasis or clarity.

Note also the difference between the **negation** of a sentence, as in *It is not dark*, and the **denial** of a sentence, as in *It is not the case that it is dark*. Denial, like its opposite, namely **affirmation**, is a mood that can be added to a sentence with a negative assertion as a result.

1.4 The connectives

Consider the sentence: *If the wind is hard, it is forbidden to swim*. Its immediate components are two complete declarative sentences *The wind is hard* and *It is forbidden to swim*. These are combined into a **conditional** sentence with the overall structure: **If**..., **then**.... The word *then* did not occur in the original but is often added in a logical context, to make the structure of conditional sentences clear. Similarly to the conditional, the two components can be combined into the sentences: *The wind is hard* **and** *it is forbidden to swim*, *The wind is hard* **or** *it is forbidden to swim*, *The wind is* **not** *hard*. The combinators in boldface are called **connectives**. We choose a basic stock of connectives and give them names and symbols. For brevity, let the letter A stand for *The wind is hard* and B for *It is forbidden to swim*:

Table 1.1 The propositional connectives

$A \& B$	$A \vee B$	$A \supset B$	$\neg A$

$A \& B$ is the **conjunction** of A and B, to be read A *and* B. $A \vee B$ is the **disjunction** of A and B, to be read A *or* B. $A \supset B$ is the **implication** of A and B, to be read *If* A, *then* B. Finally, $\neg A$ is the **negation** of A, to be read *Not* A.

The sentence A *or* B can be ambiguous: Sometimes A *or* B *or both* is meant, sometimes it is a choice between exclusive alternatives. In propositional logic, disjunction is meant in the inclusive sense, the one that is sometimes written *and/or.*

Further propositional connectives include **equivalence**: A *if and only if* B. The symbolic notation is $A \supset\subset B$. However, the four connectives of Table 1.1 will suffice for us, because other connectives can be defined in terms of them.

The symbolic notation is useful for keeping in mind that the meanings of the logical connectives are fixed and do not depend on the interpretation of a linguistic context by a user of language. The choice of symbols is historical: Most of it comes from Giuseppe Peano in the 1890s, some from Bertrand Russell in the early twentieth century, some later. The implication symbol was originally an inverted letter C, to indicate consequence. When a page was set in a printing office, the letter could be easily inverted and thus the stylized symbol $\supset$ evolved. Conjunction is found on a typewriter keyboard and the capital V disjunction symbol comes from the Latin word *vel* which means and/or. (Latin has also a word for an exclusive disjunction, namely *aut.*) The minus-sign was used for negation.

Logicians after Peano and Russell have made their own choices of symbols. Here is a partial list of symbols that have been used:

Table 1.2 Notational variants of the connectives

Conjunction:	$A \& B,\ A \wedge B,\ A \cdot B,\ AB.$
Disjunction:	$A \vee B,\ A + B,\ A \vert B.$
Implication:	$A \supset B,\ A \rightarrow B,\ A \Rightarrow B.$
Equivalence:	$A \supset\subset B,\ A \leftrightarrow B,\ A \Leftrightarrow B,\ A \equiv B.$
Negation:	$\neg A,\ -A,\ \sim A,\ \overline{A}.$

When symbolic languages were created they were sometimes accompanied by ideas about a universal language, such as Peano's creation he called

Interlingua. Other similar languages such as Esperanto have been created with the idea of promoting understanding: With a language common to all mankind, wars would end, etc. It is good to remember that part of the motivation for the development of logical languages came from such idealistic endeavours.

The one who contributed most to the development of the basic logical systems, namely Gottlob Frege, called his logical language *Begriffsschrift*, conceptual notation. He added with obvious pride that in it, 'everything necessary for a correct inference is expressed in full, but what is not necessary is generally not indicated; *nothing is left to guesswork*'.

1.5 Grammatical variation, unique readability

(a) Grammatical variation. The two sentences *If A, then B* and *B if A* seem to express the same thing. Natural language seems to have a host of ways of expressing a conditional sentence that is written $A \supset B$ in the logical notation. Consider the following list:

From A, B follows. A is a sufficient condition for B. A entails B. A implies B. B provided that A. B is a necessary condition for A. A only if B.

The last two require some thought. The **equivalence** of A and B, $A \supset\subset B$ in logical notation, can be read as *A if and only if B*, also *A is a necessary and sufficient condition for B*. Sufficiency of a condition as well as the 'if' direction being clear, the remaining direction is the opposite one. So *A only if B* means $A \supset B$ and so does *B is a necessary condition for A*.

It sounds a bit strange to say that *B is a necessary condition for A* means $A \supset B$. When one thinks of conditions as in $A \supset B$, usually A would be a cause of B in some sense or other, and causes must precede their effects. A necessary condition is instead something that necessarily follows, therefore not a condition in the causal sense.

The conjunction *A and B* in natural language can contain shades of meaning not possessed by the conjunction of propositional logic. In the sentence *John is married and his wife is Mary*, the second conjunct presupposes the first one, as can be seen by considering the sentence with the conjuncts reversed: *John's wife is Mary and he is married.*

Grammatical variation is an aspect of natural language that renders it less monotone, but that is not an issue in logic.

(b) Unique readability. In logic, the symbols are not the essential point, but the uniqueness of meaning of sentences. Let $A, B, C, \dots$ be sentences. We call them **simple** because they do not show any logical structure. Consider the **compound** sentence $A \& B \vee C$. It is ambiguous between a conjunction of A and $B \vee C$ and a disjunction of $A \& B$ and C. To avoid the ambiguity, we use parentheses with the following rule:

Whenever a sentence is part of a longer sentence, parentheses are put around it.

This rule is the basis, but it produces lots of parentheses that make it difficult to see the structure of sentences. Therefore we simplify:

Parentheses are left out if a sentence is simple. Parentheses are left out if a conjunction or a disjunction is a component of an implication.

Following the rule about parentheses, to disambiguate our example sentence $A \& B \vee C$ we write $A \& (B \vee C)$ or $(A \& B) \vee C$, and following the simplification, $A \& B \supset C \vee D$ instead of $((A)\&(B)) \supset ((C) \vee (D))$, etc.

A text is read as it is produced, in a linear succession of letters and other symbols. In reading, words appear about which one has to suspend judgement, with the expectation that the overall structure of the sentence and the meaning of individual words in it will become clear when one proceeds. Similarly with the symbolic expressions of logic: Start reading, say, the formula:

$$(A \supset B) \supset ((B \supset C) \supset (A \supset C))$$

It happens that the overall structure is seen only after a certain point. In this example, the structure 'from the outside' is an implication with the **antecedent** $A \supset B$ and the **consequent** $(B \supset C) \supset (A \supset C)$. The structure of the formula can be represented in the following **syntax tree**, drawn with the branches pointing down:

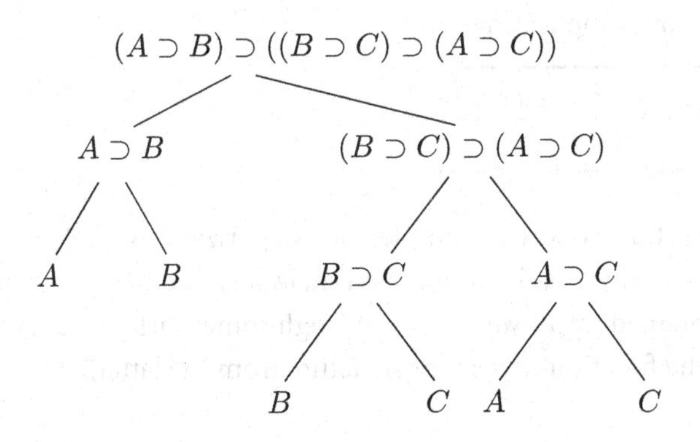

We see now how the sentence is composed of the simple sentences A, B, and C by the repeated construction of implications. There is no order in which, say, $B \supset C$ and $A \supset C$ would have been constructed. This aspect of sentences is captured by the two-dimensional tree in which these formulas do not come one before the other, but independently of each other. We could say that parentheses are used for the unique coding in a linear dimension of a thing the true nature of which is two dimensional.

We shall concentrate first on a very restricted but precise logical language, what is called **propositional logic.**

1.6 A grammar for propositional logic

Logical languages are artificial creations, but many of the standard ways of looking at natural languages apply to them. In **categorial grammar**, the basic idea about a language is that expressions have the form of a function and an argument, instead of a subject and a predicate as in traditional school grammar. An intransitive verb, say, is considered a function that is applied to an argument that must be a noun phrase, i.e., something that names an individual. Let $\mathcal{N}$ stand for the category of noun phrases and $\mathcal{S}$ for the category of sentences. In mathematics, a function f from one category of objects to another, as from $\mathcal{N}$ to $\mathcal{S}$, is categorized by the notation $f : \mathcal{N} \rightarrow \mathcal{S}$. If a is an object in the category $\mathcal{N}$, i.e., a is an expression for an individual, we categorize it through the notation $a : \mathcal{N}$. (Note that in the preceding phrase we talked first about the **object** a, then the **expression** a. It would be tedious to say all the time 'expression for an object a' instead of just a, so this qualification is left implicit.) The category of intransitive verbs is $\mathcal{N} \rightarrow \mathcal{S}$. Given an object in this category, i.e., a specific intransitive verb f, and an object a in $\mathcal{N}$, **functional application** gives us a sentence, as in the scheme:

Table 1.3 The scheme of functional application

$$\frac{f : \mathcal{N} \rightarrow \mathcal{S} \qquad a : \mathcal{N}}{f(a) : \mathcal{S}}$$

Let us make, for a concrete example, the categorizations $walk : \mathcal{N} \rightarrow \mathcal{S}$ and $John : \mathcal{N}$. Then by functional application, $walk(John) : \mathcal{S}$. The sentence *John walks* is obtained from $walk(John)$ through some 'surface transformations' that hide the function-argument structure from the latter.

We can define the language of propositional logic by giving it a categorial grammar. The category to be specified is that of **formulas**, denoted $\mathcal{F}$. We assume there to be some basic building blocks for formulas, ones that do not have any logical structure and that we call **atomic formulas** or shortly just **atoms**. These are introduced by the categorizations:

$$P : \mathcal{F}, \quad Q : \mathcal{F}, \quad R : \mathcal{F}, \ldots$$

Whenever we need to consider atoms, we can 'declare' such by writing the categorizations as above.

Following the basic idea of categorial grammar, we write the connectives as functions that are applied to formulas and that give other formulas as values. To be able to do this in a completely general way, we need letters for indicating arbitrary arguments, just like in mathematics in general one writes, say, $f(x)$, $g(y, z)$, etc. Here the variables x, y, z are just characters (letters) and nothing more. When a function is applied, a value is fed in the place of the variable that indicates the argument place or places of the function. In a similar way, we build up formulas through connectives and indicate through the letters A, B, C, ... the argument places in which specific formulas can be fed. One more thing has to be noted before we are ready to categorize the connectives: In $f(x)$, $g(y, z)$ above we had, obviously, a 'one-place' function f and a 'two-place' function g. If we consider $g(y, z)$ an abbreviation for $g(y)(z)$, we have just one-place functions. With f, we apply it to x by writing $f(x)$. With g, we apply it to y by writing $g(y)$. The result is not simply a value, but a one-place function that takes the argument z, with the writing $g(y)(z)$. Now the connectives:

Table 1.4 Connectives defined through a categorial grammar

$\dfrac{A : \mathcal{F} \quad B : \mathcal{F}}{\&(A)(B) : \mathcal{F}}$	$\dfrac{A : \mathcal{F} \quad B : \mathcal{F}}{\vee(A)(B) : \mathcal{F}}$	$\dfrac{A : \mathcal{F} \quad B : \mathcal{F}}{\supset(A)(B) : \mathcal{F}}$	$\dfrac{A : \mathcal{F}}{\neg(A) : \mathcal{F}}$

The expression $A \,\&\, B$ is obtained from the functional form $\&(A)(B)$ through the 'infix' notation for &, i.e., through writing the operation between the arguments, and by dropping the parentheses.

We note that categorial grammar automatically produces unambiguous formulas. For example, $A \,\&\, (B \vee C)$ and $(A \,\&\, B) \vee C$ are obtained through the application of the functions & and $\vee$ in different orders:

$$\dfrac{A : \mathcal{F} \quad \dfrac{B : \mathcal{F} \quad C : \mathcal{F}}{\vee(B)(C) : \mathcal{F}}}{\&(A)(\vee(B)(C)) : \mathcal{F}} \qquad \dfrac{\dfrac{A : \mathcal{F} \quad B : \mathcal{F}}{\&(A)(B) : \mathcal{F}} \quad C : \mathcal{F}}{\vee(\&(A)(B))(C) : \mathcal{F}}$$

1.7 Idealization

Our language of propositional logic has just four connectives by which the expressions are built, starting from some given simple sentences. This seems extremely restricted as compared to any natural use of a language. On the other hand, idealization is the starting point of many great discoveries. Consider the historically most successful scientific theory, namely Newton's celestial mechanics, based on Newton's three laws of motion. The simplest case to consider is the motion of two mass points. It follows from Newton's laws that if one of the points is considered fixed, the other one will move along an elliptic orbit. This result comes from an explicit solution of the Newtonian equations of motion for two mass points. Already in the case of three mass points in an otherwise empty universe, no general solution of the equations has been found. The laws have instead been applied as follows: First take the Sun, which is very massive relative to the planets. Then consider one massive planet and solve the corresponding equation to get its trajectory. Now add another planet and try to determine how much it 'perturbs' the motion of the first planet, and so on. The method is approximative and would not answer to a question such as what happens to the system if left to itself for an indefinite time. Will some planets escape in the universe, will some eventually hit the Sun, etc.? These are theoretical questions that cannot be answered by approximative methods. Just as in celestial mechanics, so also in logic: Our starting point has to be as simple as possible if we are to expect any precise results, and so we start with the forms of logical reasoning for the four propositional connectives *and, or, if..., then..., not.*

2 Rules of proof

Logical reasoning proceeds from given assumptions to some sought conclusion. The essence of assumptions is that they are hypothetical so that it is not determined if they hold, and the point with the steps of reasoning is that they produce correct conclusions whenever the assumptions are correct. These steps are two-fold: In one direction, we analyse the assumptions into their simpler parts, in another direction, we look at the conditions from which the sought for conclusion can be synthesized. The aim is to make these ends meet. Some examples lead us to a small collection of basic steps and it turns out that all logical arguments based on the connectives can be reproduced as combinations of the basic steps.

2.1 Steps in proofs

Consider our bather in Cap Breton. The argument was: We have assumptions of the forms $A \supset B$ and $\neg B$. Now A is added to these assumptions, and a contradiction follows. The argument can be presented as a succession of steps each one of which is in itself hard to doubt. We write the steps one after another together with a justification at right:

Example argument 2.1. Proof of a contradiction from $A \supset B, \neg B$, and A.

1. $A \supset B$	by assumption
2. $\neg B$	by assumption
3. A	assumed with the aim of proving a contradiction
4. B	from 1 and 3
5. $B \& \neg B$	from 4 and 2

This was simple. The conclusion on line 5 depends on the assumptions made on lines 1–3. These form the **open assumptions** of the argument.

Consider next the formula

$$(A \supset B) \supset ((B \supset C) \supset (A \supset C))$$

We encountered it already above. It is in fact correct with no assumptions made, so a general logical conclusion. To see that this is the case, assume $A \supset B$. The task is now to argue for the implication $(B \supset C) \supset (A \supset C)$ under the added assumption $A \supset B$. Assume therefore next $B \supset C$, and the task is to argue for $A \supset C$ under the assumptions $A \supset B$ and $B \supset C$. Assume therefore A, and the task is to argue for C under the assumptions $A \supset B$, $B \supset C$, and A. From $A \supset B$ and A, we get B. From $B \supset C$ and B, we get C. Therefore $A \supset C$ follows from $A \supset B$ and $B \supset C$. Therefore $(B \supset C) \supset (A \supset C)$ follows from $A \supset B$. Therefore $(A \supset B) \supset ((B \supset C) \supset (A \supset C))$ follows with no open assumptions left, i.e., as a general logical law. The individual steps in this proof can be reproduced as follows:

Example argument 2.2. A proof with no assumptions: the general logical law $(A \supset B) \supset ((B \supset C) \supset (A \supset C))$.

1. $A \supset B$ temporary assumption for proving $(B \supset C) \supset (A \supset C)$
2. $B \supset C$ temporary assumption for proving $A \supset C$
3. A temporary assumption for proving C
4. B from 1 and 3
5. C from 2 and 4
6. $A \supset C$ from 3 to 5
7. $(B \supset C) \supset (A \supset C)$ from 2 to 6
8. $(A \supset B) \supset ((B \supset C) \supset (A \supset C))$ from 1 to 7

Eight single steps were taken to make the result evident. Lines 4–5 depend on the open assumptions of lines 1–3. The formula on line 6 does not depend on the assumption on line 3, because the latter has been **closed**, and similarly for lines 7 and 8.

Let us take a more complicated example, an argument the correctness of which is not immediate. Let there be given the assumptions $A \supset B \vee C$, $B \supset D$, $C \supset E$, and $\neg E$. From these assumptions, $A \supset D$ can be concluded. To this end, we assume A to be the case. From $A \supset B \vee C$ and A, we get now $B \vee C$. There are thus two **cases**. In the second case, that of C, we get from $C \supset E$ and C the conclusion E. By the assumption $\neg E$, we get as in the bather argument $E \& \neg E$, a contradiction. Therefore case C turned out impossible and we conclude $\neg C$, so only the first case B remains. This one leads to D by the assumption $B \supset D$. Since the assumption of A has now led to D, the implication $A \supset D$ can be concluded and the temporary assumption A removed from the list of open assumptions. The individual steps are given in:

Example argument 2.3. A proof of $A \supset D$ from the four assumptions $A \supset B \vee C$, $B \supset D$, $C \supset E$, and $\neg E$.

1.	$A \supset B \vee C$	by assumption
2.	$B \supset D$	by assumption
3.	$C \supset E$	by assumption
4.	$\neg E$	by assumption
5.	A	assumed temporarily for proving D
6.	$B \vee C$	from 1 and 5
7.	C	consideration of the second case from 6
8.	E	from 3 and 7
9.	$E \& \neg E$	from 8 and 4
10.	$\neg C$	from 7 to 9
11.	B	consideration of the first case from 6
12.	D	from 2 and 11
13.	$A \supset D$	from 5 to 12

There was a good number of steps but none of them was difficult. In the handling of the two cases B and C, we could also have inferred to $B \vee C$ as on line 6, then to prove $\neg C$ by assuming C and showing that it is impossible, as on lines 7–10. Now we have $B \vee C$ and $\neg C$ and these give B as a conclusion, not as a case as on line 11.

As a last informal example, consider the following assumption: *Sentences in Thailand are twice as hard as in Finland.* The following seems to be a consequence of the assumption: *If one commits a crime in Thailand that gives a thirty-year sentence, it follows that if one commits the same crime in Finland, it gives a fifteen-year sentence.* We can formalize the latter as $(A \supset B) \supset (C \supset D)$. Now consider *If one commits a crime in Thailand that gives a thirty-year sentence and one commits the same crime in Finland, it gives a fifteen-year sentence.* We can formalize this as $(A \supset B) \& C \supset D$. It follows logically from the former:

Example argument 2.4. A proof of $(A \supset B) \& C \supset D$ from the assumption $(A \supset B) \supset (C \supset D)$.

1.	$(A \supset B) \supset (C \supset D)$	by assumption
2.	$(A \supset B) \& C$	assumed temporarily for proving D
3.	$A \supset B$	from 2
4.	$C \supset D$	from 1 and 3
5.	C	from 2
6.	D	from 4 and 5
7.	$(A \supset B) \& C \supset D$	from 2 to 6

Looking at the argument, we notice that the internal structure of $A \supset B$ was never used. Exactly the same form of argument gives in fact the conclusion $A \& B \supset C$ from the assumption $A \supset (B \supset C)$.

Could we go on with ever new examples of correct logical arguments as the assumptions grow in number and complexity? If so, what are they based on? In other words, could it happen that the modes of correct logical inference are inexhaustible?

The following turns out to be the case for the logical languages dealt with in this book, namely propositional logic and predicate logic:

There is a collection of basic steps of logical inference, such that any logical inference can be given as a suitable combination of the basic ones.

The collection of basic steps is finite and not very numerous at all, which is remarkable because *a priori*, nothing guarantees that all logically sound arguments can be coded in a finite system of rules of proof. Historically speaking, it seems not to have occurred to anyone that such is possible before Frege in 1879.

It is no fantasy to contemplate an unlimited horizon of ever new forms of correct arguments; on the contrary, such is precisely the situation in most parts of mathematics, by what is known as Gödel's incompleteness theorem for formal systems of arithmetical proof.

2.2 Negation

In example argument 2.1, we ended in a contradiction $B \& \neg B$. In example argument 2.3, case C from line 7 led to the contradiction $E \& \neg E$ on line 9, by which $\neg C$ was concluded on line 10. The lines that followed, 11–13, do not depend on line 7. This will become clear through the treatment of negation as a defined connective.

We assume there to be a sentence that is always false, called **falsity** and denoted $\perp$. It can be considered from the categorial grammar point of view a connective with zero number of arguments. Thus, we categorize it by $\perp : \mathcal{F}$.

Definition 2.1. $\neg A \equiv_{df} A \supset \perp$.

The notation $\equiv_{df}$ means that the expression at left is an **abbreviation** for the expression at right. The subscript *df* is often left out from such definitions. With the definition of negation, the conclusion $\perp$ follows from the assumptions $\neg A$ and A, just like B follows from the assumptions $A \supset B$ and A.

2.3 Natural deduction in a linear form

We shall put up a system of proof for the three connectives &, $\supset$, and $\neg$. The rules have been already used informally in the example arguments of the previous section. In example argument 2.1, we concluded $B \,\&\, \neg B$ from B and $\neg B$. The rule by which the conclusion is drawn is:

Conjunction introduction: *To conclude a formula of the form $A \,\&\, B$, it is sufficient to have the components A and B.*

The components are called the **premisses** of the rule and $A \,\&\, B$ its **conclusion.** The premisses can be open assumptions or conclusions of previously applied rules. When the rule of conjunction introduction is applied, the lines of the components as well as the application of the rule are indicated as in the following:

1. A assumption
2. B assumption
3. $A \,\&\, B$ 1, 2, $\&I$

There is a similar introduction rule for implication:

Implication introduction: *To conclude a formula of the form $A \supset B$, it is sufficient to make the temporary assumption A and to arrive at B.*

One would usually arrive at B some lines down from the making of the assumption A. After $A \supset B$ has been concluded, A and whatever depends on it cannot be used. To indicate that an assumption is a temporary one, or a **hypothesis** that will be **closed** later, a stroke is drawn at the beginning of the line. When the hypothesis is closed, the stroke is completed into a bracket to exclude the use of anything on the lines that depend on the hypothesis, as in the following:

1.	A	hypothesis: goal B
$\vdots$		
n.	B	
$n+1$.	$A \supset B$	$1-n, \supset I$

If we apply later, say, rule $\&I$ to infer some formula $C \,\&\, D$, the components C and D can be on any lines that are not inside the bracket drawn for the closing of a hypothesis in any instance of rule $\supset I$.

We give as a first example of the use of the rule system a proof of the formula $A \supset (B \supset A \& B)$:

┌	1. A	hypothesis: goal $B \supset A \& B$
│┌	2. B	hypothesis: goal $A \& B$
││	3. $A \& B$	1,2, $\&I$
│└	4. $B \supset A \& B$	2–3, $\supset I$
└	5. $A \supset (B \supset A \& B)$	1–4, $\supset I$

Looking at the formula to be proved, we notice that its outermost connective is an implication. Therefore we try to conclude it by rule $\supset I$. It will be sufficient to assume the antecedent A of the implication and to prove the consequent $B \supset A \& B$. The latter is again an implication so that B can be assumed with the aim of proving $A\&B$. To prove the latter, it is by rule $\&I$ sufficient to have the conjuncts A and B available separately, which is the case. Lines 3–5 record the steps that are made to arrive at the goals indicated in the making of the hypotheses.

If an assumption is of the form $A \& B$, it can be split into its components A and B. These steps are represented by the two rules:

Conjunction elimination: *From $A \& B$, A and B can be concluded.*

These rules are used as in the following two schemes:

1. $A \& B$		1. $A \& B$	
2. A	1, $\&E_1$	2. B	1, $\&E_2$

The conclusion need not be on the next line but can be drawn whenever it is needed, provided that the premiss $A \& B$ is not inside a closed bracket at that point. We often drop the subscripts from $\&E_1$, $\&E_2$ because it can be seen from the premiss and the conclusion which of the two variants of the rule was used.

The rule of implication elimination has been used in all of the example arguments of Section 2.1:

Implication elimination: *From $A \supset B$ and A together, B can be concluded.*

The rule is used as in the following scheme:

1. $A \supset B$	
2. A	
3. B	1, 2, $\supset E$

Again, the conclusion can occur at any later line, as long as the premisses are not inside closed brackets.

As a last rule for the connectives &, $\supset$, and $\bot$, we have

Falsity elimination: *From* $\bot$, *any formula can be concluded.*

The rule is often called by its Latin name, *ex falso quodlibet*, from a falsity anything follows.

1. $\bot$
2. C $1, \bot E$

There is no introduction rule for falsity.

The system of rules for &, $\supset$, and $\bot$ is called a system of **natural deduction**, after Gerhard Gentzen who invented it in 1932. Earlier systems of logic were based on **logical axioms** and rule $\supset E$, called after its Latin name *modus ponens* (something like 'the mode of positing'). The idea in axiomatic systems of logic was that logical axioms express basic **logical truths**, and axioms are used for deriving other logical truths. One drawback of the approach is that some of the axioms are rather complicated, perhaps not recognized as logical truths, whereas some of the theorems are among the simplest. For example, the formula $A \supset A$ is a theorem with a proof in axiomatic logic that is not easy to find. Gentzen's system, in contrast, is natural in the sense that it corresponds to the way in which proofs are actually made in the deductive sciences. Proofs have a very clear structure and are as a rule easy to find.

We can now give by Gentzen's rules of natural deduction a formal derivation of example argument 2.2:

1.	$A \supset B$	hypothesis: goal $(B \supset C) \supset (A \supset C)$
2.	$B \supset C$	hypothesis: goal $A \supset C$
3.	A	hypothesis: goal C
4.	B	$1, 3, \supset E$
5.	C	$2, 4, \supset E$
6.	$A \supset C$	$3\text{–}5, \supset I$
7.	$(B \supset C) \supset (A \supset C)$	$2\text{–}6, \supset I$
8.	$(A \supset B) \supset ((B \supset C) \supset (A \supset C))$	$1\text{–}7, \supset I$

2.4 The notion of a derivation

When the sentences of logic are defined precisely, as in Section 1.6, they are called **formulas**. Similarly, when the principles of proof are made explicit, proofs that follow these rules are called **derivations**. The last line in a derivation is the **conclusion**. Assumptions, possibly none, that have not been closed, are the **open assumptions** of a derivation.

Definition 2.2. *Formula B is* **derivable** *from the assumptions* $A_1, \dots A_n$ *if there is a derivation with B as a conclusion and* $A_1, \dots A_n$ *as open assumptions.*

We shall write $A_1, \dots, A_n \vdash B$ to indicate derivability. The symbol $\vdash$ denotes the **derivability relation** between assumptions, possibly none, and a conclusion. The last line of a derivation is also called its **endformula**. If there are no open assumptions left in a derivation of B, i.e., if $\vdash B$, then B is a **theorem**. Theorems give general logical laws that are correct with no assumptions made.

There is a limiting case of a derivation in which an assumption A is made. It is at the same time a derivation of the conclusion A from the assumption A, as in:

1. A	hypothesis: goal A
2. $A \supset A$	$1, \supset I$

In terms of the derivability relation, the hypothesis on line 1 can be written as $A \vdash A$ and line 2 as $\vdash A \supset A$. Consider as another case $A \supset (B \supset A)$. Verbally, if we assume A, then A follows under any other assumption B:

1. A	hypothesis: goal $B \supset A$
2. $B \supset A$	$1, \supset I$
3. $A \supset (B \supset A)$	$1\text{–}2, \supset I$

This does not look particularly nice: We have closed an assumption B that was not made. But if we say that an assumption was used 0 times, the thing starts looking more reasonable. Consider the opposite case, one in which an assumption is used several times, as in the following derivation of the logical law $(A \supset ((A \supset B)) \supset (A \supset B)$:

1. $A \supset (A \supset B)$	hypothesis: goal $A \supset B$
2. A	hypothesis: goal B
3. $A \supset B$	$1, 2, \supset E$
4. B	$3, 2, \supset E$
5. $A \supset B$	$2\text{–}4, \supset I$
6. $(A \supset (A \supset B)) \supset (A \supset B)$	$1\text{–}5, \supset I$

Note the order in which the premisses are listed on line 4: First the **major premiss** with the connective, from line 3, then the **minor premiss** from line

2. Note further that assumption A from line 2 was used twice in $\supset E$, on line 3 and on line 4. It is known that the endformula of the derivation cannot be derived without such **multiple** use of an assumption.

Analogously to a multiple use of an assumption, we can say that assumption B in the derivation of $A \supset (B \supset A)$ was used **vacuously**. Another way in which a formula has to be used twice is in the derivation of $A \supset A \& A$:

1. A	hypothesis: goal $A \& A$
2. $A \& A$	1, 1, $\& I$
3. $A \supset A \& A$	1–2, $\supset I$

The components of the conjunction came from the same line. The next chapter gives Gentzen's original system of natural deduction in which such unniceties will not occur.

We summarize the system of rules of natural deduction for &, $\supset$, and $\bot$:

Table 2.1 Rules of linear natural deduction

$\& I$:	*Premisses A and B, conclusion A&B.*
$\supset I$:	*Premiss B derived from the hypothesis A, conclusion $A \supset B$.*
$\& E$:	*Premiss $A \& B$, conclusion A. Premiss $A \& B$, conclusion B.*
$\supset E$:	*Premisses $A \supset B$ and A, conclusion B.*
$\bot E$:	*Premiss $\bot$, conclusion C.*

Compared to Gentzen's natural deduction that uses a **tree form**, the formulas are arranged in a linear succession, one after the other, and the construction of derivations is straightforward.

2.5 How to construct derivations?

The task is to show that a conclusion B can be reached from the assumptions $A_1, \ldots, A_n$. The following 'manual' will be helpful in finding a way from given assumptions to a sought conclusion.

Table 2.2 Procedure for finding derivations in linear natural deduction

1. *Write down the assumptions $A_1, \ldots, A_n$ as the first n lines of a derivation.*
2. *Check if any of the assumptions can be analysed into components by the E-rules.*
3. *When no E-rule applies, check from the conclusion what I-rule can be used to derive it and try to derive the premisses of that rule.*
4. *Repeat 2 and 3 until the derivation succeeds.*

These instructions will give a derivation in most cases, but there are other cases in which a dead end is reached and one has to have some insight into how to proceed. The following examples are straightforward:

a. $(A \supset B) \& (B \supset C) \vdash A \supset C$

1. $(A \supset B) \& (B \supset C)$	assumption
2. $A \supset B$	1, $\&E$
3. $B \supset C$	1, $\&E$
4. A	hypothesis: goal C
5. B	2, 4, $\supset E$
6. C	3, 5, $\supset E$
7. $A \supset C$	4–6, $\supset I$

b. $A \supset B \vdash \neg(A \& \neg B)$

1. $A \supset B$	assumption
2. $A \& \neg B$	hypothesis: goal $\bot$
3. A	2, $\&E$
4. $\neg B$	2, $\&E$
5. B	1, 3, $\supset E$
6. $\bot$	4, 5, $\supset E$
7. $\neg(A \& \neg B)$	2–6, $\supset I$

Remember that negation is a special case of implication, so that line 6 in this derivation is concluded by $\supset E$ and line 7 by $\supset I$.

c. $\vdash A \& B \supset \neg\neg A \& \neg\neg B$.

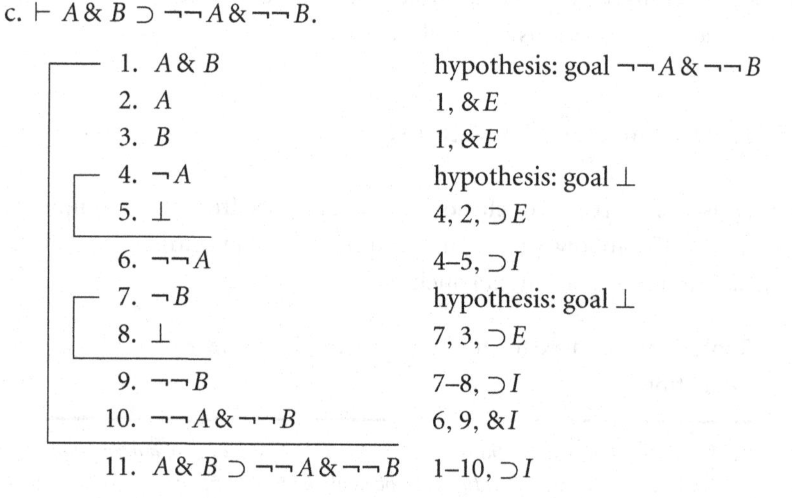

1. $A \& B$	hypothesis: goal $\neg\neg A \& \neg\neg B$
2. A	1, $\&E$
3. B	1, $\&E$
4. $\neg A$	hypothesis: goal $\bot$
5. $\bot$	4, 2, $\supset E$
6. $\neg\neg A$	4–5, $\supset I$
7. $\neg B$	hypothesis: goal $\bot$
8. $\bot$	7, 3, $\supset E$
9. $\neg\neg B$	7–8, $\supset I$
10. $\neg\neg A \& \neg\neg B$	6, 9, $\&I$
11. $A \& B \supset \neg\neg A \& \neg\neg B$	1–10, $\supset I$

Eleven steps were required to check that **double negations** can be added in front of conjuncts. None of the steps required more than a following of the instructions for constructing derivations. Three temporary assumptions were made in the derivation. These hypotheses were closed in an order that comes from the following principle:

Closing of hypotheses. *In a derivation, the hypothesis made last must be closed first.*

In graphical terms, the effect is that the brackets do not cross each other. They either follow one another or appear one completely inside another. This rule will be sufficient for the correct handling of hypotheses.

One could think that such detail as in the last derivation is somehow exaggerated. The point is that each step is as simple as possible, and that complexity comes from the putting together of the elementary steps. When the syntax of the derivation trees is specified in the same detail as that of the formulas in Section 1.6, the correctness of a derivation can be checked by a program. Errors in the construction of derivations are just errors in syntax, i.e., basically the same as errors in writing.

2.6 Schematic formulas, schematic derivations

As explained in Section 1.6, the sentences of propositional logic are built from some given atomic sentences $P, Q, R, \dots$ and $\bot$ through the use of connectives. The letters $A, B, C, \dots$ that we use are **schematic**. We indicate by them **arbitrary** sentences, just like one indicates arbitrary numbers by $a, b, c, \dots$ in arithmetical laws such as $a + b = b + a$. Each **instance** of such a law has some concrete numbers in place of the schematic letters, such as in $7 + 5 = 5 + 7$.

Consider the sentence

$$(A \,\&\, B) \,\&\, C \supset D$$

It is not really a sentence, because sentences are built from atomic sentences and falsity. It just gives a possible form of a sentence, although we would not usually say 'form of a sentence' but just 'sentence' for short. Let the following four atomic sentences be **substituted** for A, B, C, and D in the schematic sentence:

Point a is incident on line l,
Point b is incident on line l,
Points a and b are distinct,
The connecting line of points a and b is equal to the line l.

The result of substitution is the sentence:

(*Point a is incident on line l* & *Point b is incident on line l*) & *Points a and b are distinct* $\supset$ *The connecting line of points a and b is equal to the line l.*

It goes in the same way for derivations: The derivations we have constructed are schematic, because the formulas in them are schematic, as in:

1. A	hypothesis: goal $\neg\neg A$
2. $\neg A$	hypothesis: goal $\bot$
3. $\bot$	2, 1, $\supset E$
4. $\neg\neg A$	2–3, $\supset I$
5. $A \supset \neg\neg A$	1–4, $\supset I$

Any derivation in which A is a concrete sentence is a correct derivation of a concrete instance of $A \supset \neg\neg A$. Say, if we substitute in the whole derivation for A the sentence *Point a is incident with line l*, we get a derivation of:

Point a is incident with line l $\supset \neg\neg$ *Point a is incident with line l*

What if the arbitrary sentence A in the above derivation is a negation, say, of the form $\neg B$? Substitution of $\neg B$ for A in the schematic derivation will give another schematic derivation:

1. $\neg B$	hypothesis: goal $\neg\neg\neg B$
2. $\neg\neg B$	hypothesis: goal $\bot$
3. $\bot$	2, 1, $\supset E$
4. $\neg\neg\neg B$	2–3, $\supset I$
5. $\neg B \supset \neg\neg\neg B$	1–4, $\supset I$

The replacement was done by an algorithm (a human writer would have perfected the job by aligning the justifications). This text is produced by the LaTeX text editing program. It has a command by which any string of symbols in the source file for a text can be replaced by any other string of symbols. The printed version of the derivation was compiled from the source file that is visible on the screen in a way that emulates the style of writing with a traditional typing machine: Each letter has a constant width, each letter looks the same, etc. Things such as *italics* or logical symbols are produced by special commands. The `typewriter style` on the screen can be reproduced in the printed text. The substitution of A by $\neg B$ was done so that a command was given for replacing the string of symbols `A` in the source file for the derivation by the string of symbols `\neg B`. In the same way, we can replace A in the derivation by $\neg A$, to get a derivation of:

$$\neg A \supset \neg\neg\neg A$$

We can make a **simultaneous** substitution of several sentences in a formula or a derivation. The equivalence $A \,\&\, B \supset\subset B \,\&\, A$ consists of the conjunction of the two implications $A \,\&\, B \supset B \,\&\, A$ and $B \,\&\, A \supset A \,\&\, B$. When we have derived $A \,\&\, B \supset B \,\&\, A$, we obtain from that derivation a derivation of $B \,\&\, A \supset A \,\&\, B$ by substituting A for B and B for A. It has to be simultaneous: If we first substituted B for A in $A \,\&\, B \supset B \,\&\, A$, we would get $B \,\&\, B \supset B \,\&\, B$. If next we substituted A for B in the latter, we would get $A \,\&\, A \supset A \,\&\, A$, not $B \,\&\, A \supset A \,\&\, B$ as in the simultaneous substitution. Once A is replaced by B in $A \,\&\, B \supset B \,\&\, A$, the structure of the original file is messed up: There is no way of recovering it from the result of the replacement.

A simultaneous substitution is in practice done as a sequence of single substitutions. In the example, we can avoid mixing occurrences of A and B by choosing new symbols C and D, then substituting first C for A, then D for B, then A for D, and last B for C.

What symbols we use is of no consequence, as long as the form of sentences and derivations is not affected.

2.7 The structure of derivations

A logician would not be interested in the mere making of formal derivations, but in more general questions. How can the structure of derivations in a given system of rules be characterized? Is there a method for automatically producing derivations? What if some detail of the system of rules is changed?

A first general property of a system of rules is that it does not lead to a contradiction, i.e., that the formula $\bot$ is not derivable in the system. It is said that the system is then **consistent**. It is not such a trivial task to establish consistency as might seem at first, because the number of possible derivations has no bound. So one has to prove that none of the infinitely many derivations that leave no open assumptions cannot have the endformula $\bot$.

If we look at the derivations in Section 2.5, especially the order of application of the rules, we notice that after the assumptions have been made, elimination rules are applied. Towards the end of a derivation, there are instead applications of introduction rules. Example 2.5c deviates somewhat from this pattern, because new hypotheses appear after some eliminations have been made. This phenomenon is caused by the linear arrangement of formulas.

Consider now a derivation that has a part such as:

$$
\begin{array}{lll}
\vdots & & \\
n. & A & \\
\vdots & & \\
m. & B & \\
\vdots & & \\
k. & A \,\&\, B & n, m, \&I \\
\vdots & & \\
l. & A & k, \&E \\
\vdots & &
\end{array}
$$

Assume the derivation to be such that the open assumptions on which the occurrence of formula A on line l depends are all included among the open assumptions on which the occurrence of formula A on line n depends. Then the derivation has a **loop**, or **cycle**, because the same formula appears twice and no new assumptions have been made. We can delete the part of derivation between the two occurrences, to get a simpler derivation:

$$
\begin{array}{ll}
\vdots & \\
n. & A \\
\vdots &
\end{array}
$$

The derivation continues from line n as the original derivation continued from line l. Also here the linear arrangement of formulas in derivations can be problematic: What if some formulas from the part that is cut out are used to justify steps after line l? How to fix it?

We notice that a conjunction with the components A and B was first introduced, then eliminated, against the standard order of eliminations followed by introductions in the examples of the previous section. The step of elimination after a step of introduction makes the second conjunct B a useless part of the derivation and that part can be deleted. By this deletion, the standard order is re-established.

The main result about systems of natural deduction is that all parts of derivations in which an introduction is followed by a corresponding elimination on the introduced formula, can be removed from a derivation. It must be warned again that the line of the 'corresponding elimination' is not fixed in the linear variant of natural deduction, which complicates things. The net effect of removing introductions followed by eliminations is a derivation in **normal form**, a notion to be defined in a precise way in the next chapter.

Notes and exercises to Chapter 2

The arrangement of derivations in a linear succession is often called the 'Fitch-system' of natural deduction, after its appearance in the book *Symbolic Logic* by Frederic Fitch (1952). We have followed the handsome bracket notation of the Swedish textbook Prawitz (1991).

1. Give a statement in words of the formula on line 8 of example argument 2.2.

2. Show by an informal argument that $A \& B \supset C$ follows from the assumption $A \supset (B \supset C)$.

3. Give derivations of the following:
   a. $\vdash A \& B \supset A$
   b. $\vdash (A \supset B) \& A \supset B$
   c. $\vdash (A \supset B) \& (B \supset C) \supset (A \supset C)$
   d. $\vdash A \& \neg A \supset \bot$
   e. $\vdash (A \supset B) \supset (\neg B \supset \neg A)$
   f. $\vdash A \supset \neg\neg A$
   g. $\vdash (A \supset (B \supset C)) \supset (B \supset (A \supset C))$
   h. $\vdash (A \& B \supset C) \supset (A \supset (B \supset C))$

4. More derivations:
   a. $\vdash \neg\neg\neg A \supset \neg A$
   b. $\vdash (A \supset \neg B) \supset (B \supset \neg A)$

c. $\neg(A \& \neg B) \vdash A \supset \neg\neg B$
d. $A \& \neg B \vdash \neg(A \supset B)$
e. $(A \supset B) \& (A \supset C) \vdash A \supset B \& C$
f. $A \& B \supset C \vdash B \& A \supset C$
g. $A \supset B \& C \vdash A \supset C \& B$

3 Natural deduction

The linear variety of natural deduction makes it possible to construct derivations in steps, one after the other. On the other hand, we have not treated disjunction yet, and we have noticed that the normal form of derivations would not be transparent and simple in a linear arrangement of formulas. Both of these defects are corrected when we now turn to a study of Gentzen's original system of natural deduction for propositional logic. Formulas in derivations are arranged in a **tree form**, such that each formula is either an assumption or the conclusion of exactly one logical rule, and each formula except the endformula of the whole derivation is a premiss of exactly one logical rule. When we here talk about 'each formula', we mean more precisely each single **formula occurrence** in a rule instance in a derivation, but don't repeat that each time.

Tree derivations were in practice a novelty with Gentzen and their widespread use in logic derives from his doctoral thesis (1934–5). He took the idea over from the work of Paul Hertz of the 1920s. The tree form shows 'what depends on what' in a derivation and makes it possible to transform the order of application of rules; the most central methodological novelty in Gentzen that soon led to spectacular results about the structure of proofs.

The tree form has a problematic aspect, namely, that a tree would grow naturally from its root to the leaves at the ends of the branches, but trees in natural deduction are constructed in the wrong way. We have to start from the leaves, keep things in mind, and try to fit it all together.

3.1 From linear derivations to derivation trees

The following instructions will automatically produce derivations in the original tree form of Gentzen from the linear derivations of Chapter 2:

Table 3.1 Translation of linear derivations to tree form

---

1. *Write down the endformula and a line above it.*
2. *Write next to the line the rule that was used in concluding the endformula. If it was* $\supset I$*, write after the rule the number of the line on which the hypothesis closed by the rule occurred.*
3. *Write above the line, from left to right, the formula or formulas that correspond to the numbers that justified the application of the rule. If it was* $\supset I$*, write the consequent of the implication.*
4. *Repeat the above until you come along each branch of the derivation tree to an assumption. If it is temporary, i.e., a hypothesis, write the number of its line above it.*

The following derivations are produced when these instructions are applied to example derivations (a)–(c) of Section 2.5:

a. $(A \supset B) \& (B \supset C) \vdash A \supset C$

$$\dfrac{\dfrac{\dfrac{(A \supset B) \& (B \supset C)}{B \supset C}\,{\scriptstyle \& E_2} \qquad \dfrac{\dfrac{(A \supset B) \& (B \supset C)}{A \supset B}\,{\scriptstyle \& E_1} \qquad \overset{4}{A}}{B}\,{\scriptstyle \supset E}}{C}\,{\scriptstyle \supset E}}{A \supset C}\,{\scriptstyle \supset I,4}$$

The number 4 after the downmost **inference line** and rule symbol $\supset I$ is an **assumption label**, or **label** for short. The translation produces the number also above the hypothesis A, to indicate which hypothesis is closed where in a derivation. The assumption $(A \supset B) \& (B \supset C)$ of the derivation was used twice in the linear variant, and the formula appears correspondingly with two occurrences in the tree. The hypothesis A occurs as a topformula.

Note that there is no need to write, next to the rules, where their premisses come from, because we get by the translation:

In a derivation tree, the premisses of a rule stand directly above and the conclusion directly below the inference line.

b. $A \supset B \vdash \neg(A \& \neg B)$

$$\dfrac{\dfrac{\dfrac{\overset{2}{A \& \neg B}}{\neg B}\,{\scriptstyle \& E_2} \qquad \dfrac{A \supset B \qquad \dfrac{\overset{2}{A \& \neg B}}{A}\,{\scriptstyle \& E_1}}{B}\,{\scriptstyle \supset E}}{\bot}\,{\scriptstyle \supset E}}{\neg(A \& \neg B)}\,{\scriptstyle \supset I,2}$$

The closed assumption is again a topformula, contrary to the linear derivation.

c. $\vdash A \& B \supset \neg\neg A \& \neg\neg B$

$$\dfrac{\dfrac{\dfrac{\dfrac{\overset{4}{\neg A} \quad \dfrac{\overset{1}{A \& B}}{A}\,{\scriptstyle \& E_1}}{\bot}\,{\scriptstyle \supset E}}{\neg\neg A}\,{\scriptstyle \supset I,4} \qquad \dfrac{\dfrac{\overset{7}{\neg B} \quad \dfrac{\overset{1}{A \& B}}{B}\,{\scriptstyle \& E_2}}{\bot}\,{\scriptstyle \supset E}}{\neg\neg B}\,{\scriptstyle \supset I,7}}{\neg\neg A \& \neg\neg B}\,{\scriptstyle \& I}}{A \& B \supset \neg\neg A \& \neg\neg B}\,{\scriptstyle \supset I,1}$$

The linear derivation had elimination and introduction steps mixed with each other. The tree form instead has, along each **derivation branch**, assumptions followed by eliminations followed by introductions. Thus, the derivation is normal because no introduction is followed by an elimination.

Let us look next at the translation of some of the problematic derivations in Section 2.4. The translation of the linear derivation of $A \supset (B \supset A)$ is:

$$\dfrac{\dfrac{\overset{1}{A}}{B \supset A}\,{\scriptstyle \supset I}}{A \supset (B \supset A)}\,{\scriptstyle \supset I,1}$$

When we come to line 2 of the linear derivation, the translation manual tells us to write after the inference line the number of the line in the linear derivation on which the hypothesis B is made. There is no such number to be written, so only the rule is indicated.

The translation of the linear derivation of $(A \supset (A \supset B)) \supset (A \supset B)$ is left as an exercise. The derivation of $A \supset A \& A$ becomes:

$$\dfrac{\dfrac{\overset{1}{A} \quad \overset{1}{A}}{A \& A}\,{\scriptstyle \& I}}{A \supset A \& A}\,{\scriptstyle \supset I,1}$$

From the translations of this section, we notice the following:

The justifications, i.e., the references to the lines of the premisses in the steps of a linear derivation, are nothing but a codification of a two-dimensional derivation tree.

Looking at the translation of example (c), we notice that the order in which the hypotheses $\neg A$ and $\neg B$ were made in the linear derivation would not make a difference in the translated derivation. The order in which the hypotheses are treated could be chosen freely; for had we first assumed $\neg B$

and derived $\neg\neg B$, then the same with $\neg A$, the translation would have been identical save for the numbers used as labels. Therefore we have:

Different linear derivations can correspond to one and the same tree derivation.

3.2 Gentzen's rules of natural deduction

A translation of the rules of linear natural deduction into a tree form gives the following rule system:

Table 3.2 Gentzen's rules for &, $\supset$, and $\perp$

$$\frac{A \quad B}{A \,\&\, B}\,\&I \qquad \frac{A \,\&\, B}{A}\,\&E_1 \qquad \frac{A \,\&\, B}{B}\,\&E_2$$

$$\frac{\begin{matrix} \overset{1}{A} \\ \vdots \\ B \end{matrix}}{A \supset B}\,\supset I,1 \qquad \frac{A \supset B \quad A}{B}\,\supset E \qquad \frac{\perp}{C}\,\perp E$$

In the schematic rule $\supset I$, any number of occurrences of A can be closed. If this number is 0, we have a **vacuous discharge** (or **closing**) of an assumption, if more than 1, a **multiple** discharge. Otherwise a discharge is **simple**. Each instance of rule $\supset I$ must have a **fresh** label, one that has not been used in any other rule instance.

We add to the above system of introduction and elimination rules what is sometimes called 'the rule of assumption':

Table 3.3 The rule of assumption

$$A$$

This rule just means that we are free to start a derivation branch with any formula used as an assumption, temporary or permanent. The notation does not distinguish between these two kinds of assumptions, and, indeed, we need not decide, when starting a derivation, which assumptions will be closed in the end. They can all be closed by applications of rule $\supset I$, as Gentzen did in his original work.

The assumptions in a derivation tree have a well-defined **multiplicity**: We can count the number of times an assumption occurs. Take now any formula C in a derivation and proceed up from it along all possible branches. Collect the open assumptions, with their multiplicity recorded, and you have what is called the **context** of formula C in the derivation. These contexts are denoted by capital Greek letters $\Gamma, \Delta, \Theta, \Lambda, \ldots$.

Two derivations can be **composed** into one under a simple condition: Given a derivation of A from Γ and of C from A and Δ, such that all labels in the two derivations are distinct, they can be put together into a derivation of C from Γ and Δ:

Table 3.4 Composition of two derivations

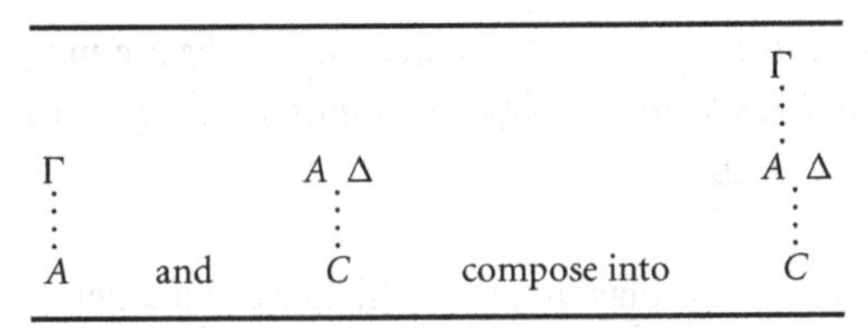

Should there be a label common to the two derivations, one of them can be changed. Composition can be presented also through the derivability relation: We have $\Gamma \vdash A$ and $A, \Delta \vdash C$ and these are composed into $\Gamma, \Delta \vdash C$.

The endformula of a derivation would often be concluded by an introduction. An assumption, on the other hand, would typically be put into use through an elimination. Therefore composition usually produces an introduction followed by an elimination, i.e., a non-normality.

The translations into tree form in the previous section are examples of derivations with Gentzen's rules of natural deduction. Here is one more:

$\vdash (A \,\&\, B \supset C) \supset (A \supset (B \supset C))$

$$\cfrac{\cfrac{\cfrac{\cfrac{\overset{3}{A \,\&\, B \supset C} \qquad \cfrac{\overset{2}{A} \quad \overset{1}{B}}{A \,\&\, B}{\scriptstyle \&I}}{C}{\scriptstyle \supset E}}{B \supset C}{\scriptstyle \supset I,1}}{A \supset (B \supset C)}{\scriptstyle \supset I,2}}{(A \,\&\, B \supset C) \supset (A \supset (B \supset C))}{\scriptstyle \supset I,3}$$

An introduction rule is followed by an elimination rule in the upper right branch, but the major premiss of the elimination is not derived by an introduction. Therefore the derivation is normal.

The construction of derivation trees has some awkward aspects: When one starts the derivation, it is not clear where the assumptions should be written and how parts of derivations fit together. Maybe the lines don't match when one comes to concluding formulas by rules that have two premisses, maybe the premisses are in a wrong order from left to right, etc. The same problem would be met if one had to draw a syntax tree for a formula, as in Section 1.5, but in a reversed order, starting from the simple formulas. To see where they have to be placed, one would have to know the whole tree. We conclude that Gentzen's tree derivation form does not support **proof search** in an optimal way. In practice, one will learn anyway to construct tree derivations.

The drawbacks in proof search are compensated by a neat property of the rules that we mentioned above:

In a tree derivation, each formula except the endformula is the premiss of the rule above which it stands. Each formula except an assumption is the conclusion of the rule under which it stands.

Thus, we do not need to use formulas that stand higher up in a derivation, nor do formulas remain 'unused' after they have been concluded. These advantages over the linear form of derivation make it possible to give a definition of a normal derivation without compromises:

Definition 3.1. Normal derivation in Gentzen's natural deduction: *A derivation in Gentzen's natural deduction for* &, $\supset$, *and* $\perp$ *is* **normal** *if no major premiss of an E-rule is derived by an I-rule.*

In a linear derivation, a formula can be introduced but a 'corresponding elimination', as we wrote in Section 2.7, can come much later, for which reason the description of normal derivations in that section has to be taken in somewhat figurative terms.

The main result about natural deduction states that derivations can be transformed into a normal form. The proof, postponed to Section 13.2, consists in defining the transformations and in showing that their application terminates in a derivation that is in a normal form. Corresponding to the rules for the connectives & and $\supset$, the **non-normalities** to be transformed are the three *I-E* pairs, called **detour convertibilities**:

Table 3.5 Detour convertibilities on & and $\supset$

$\dfrac{\dfrac{\overset{\vdots}{A}\quad\overset{\vdots}{B}}{A\,\&\,B}\,\&I}{\underset{\vdots}{A}}\,\&E_1$	$\dfrac{\dfrac{\overset{\vdots}{A}\quad\overset{\vdots}{B}}{A\,\&\,B}\,\&I}{\underset{\vdots}{B}}\,\&E_2$	$\dfrac{\dfrac{\begin{array}{c}\overset{1}{A}\\ \vdots\\ B\end{array}}{A\supset B}\supset I,1\quad \overset{\vdots}{A}}{\underset{\vdots}{B}}\supset E$

These convertibilities are eliminated by transforming the derivations into:

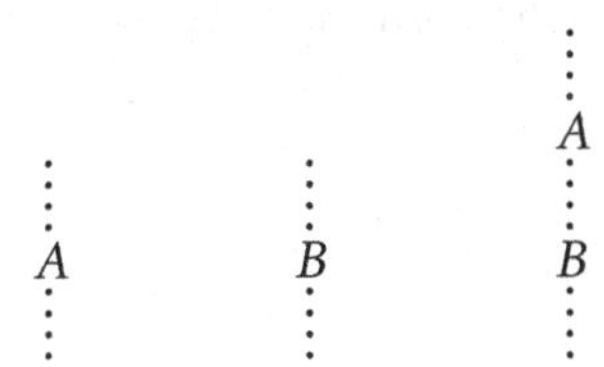

The first one is obtained by taking the derivation of the first premiss of $\&I$ and continuing through composition from the conclusion of $\&E_1$, the second analogously, and the third by taking the derivation of the minor premiss A of $\supset E$, continuing from A to B as in the derivation of the premiss of $\supset I$, then finishing by proceeding from B as in the conclusion of $\supset E$.

Given a derivation of a conclusion C from the assumptions Γ, any conversion of the derivation should produce a derivation in which there are no new assumptions. This is obviously the case with the conversions we have considered: At most some assumptions may disappear, say, those that were used to derive B in the first detour conversion above. The result of a conversion can thus be an improvement on the original derivation: the same conclusion but possibly fewer assumptions.

What if the discharge of A in rule $\supset I$ was not simple? If it was an empty discharge, there is a derivation of B without assumption A and the I-E pair can be eliminated. If it was a multiple discharge, with n copies of A, the conversion produces a more complicated result that we can depict as:

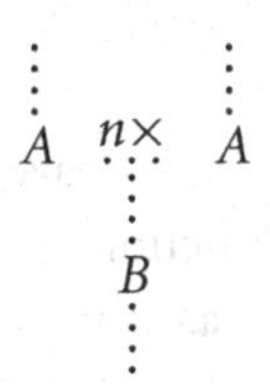

Gentzen's normalization proof is based on a measure of complexity of derivations. Each conversion lowers that complexity, until a normal derivation is reached. With a multiple discharge in rule $\supset I$, the derivation of the minor premiss gets multiplied, and the difficulty is to find a complexity measure such that a detour conversion on $\supset$ compensates the increase in complexity caused by the multiplication by n of the derivation of A.

The normal form theorem leads to the **subformula property** of normal derivations: All formulas in such derivations are parts of the open assumptions or the conclusion. In propositional logic, it then follows that the question of derivability is **decidable**: There is an **algorithm of proof**

search that either produces a sought derivation of B from given assumptions $A_1, \dots A_n$ or terminates with a failed proof search and **underivability** as a result.

3.3 Derivations with cases

It is time to add disjunction introduction and elimination to our system of rules for conjunction, implication, and negation. With this addition, we have what is called **intuitionistic propositional logic**. The term has its historical origins in the intuitionistic mathematics of L. Brouwer, a version of the constructive tendency in mathematics and developed mainly in the 1920s. By the end of this era, a general logical structure was distilled from intuitionistic mathematics, as explained in Section 14.4. It is a remarkable result that the natural rules of the logical connectives give intuitionistic logic as a result, with no conditions that would force this to be so.

The pattern for the elimination rule for disjunction is different from the previous elimination rules, which leads to a new type of simplification of derivations, called **permutative conversion.**

(a) Rules for disjunction. The introduction rules for disjunction are:

Table 3.6 The introduction rules for $\vee$

$$\frac{A}{A \vee B}\,\vee I_1 \qquad\qquad \frac{B}{A \vee B}\,\vee I_2$$

As with the two E-rules for &, we can leave unwritten the subscripts.

The elimination rule for disjunction was used in some way in example argument 2.3. We had there two **cases** and a conclusion that followed independently of which case happened to obtain. Thus, if we have a disjunction $A \vee B$, if C follows from A and if C follows from B, then C follows in any case. The rule is written as:

Table 3.7 The elimination rule for $\vee$

$$\frac{A \vee B \quad \begin{matrix} \overset{1}{A} \\ \vdots \\ C \end{matrix} \quad \begin{matrix} \overset{1}{B} \\ \vdots \\ C \end{matrix}}{C}\,\vee E,1$$

The hypotheses A and B can occur any number of times in the two sub-derivations of the minor premiss C, similarly to rule $\supset I$, with vacuous, simple, or multiple discharge in each.

Examples of derivations with cases:

a. $A \vee B \vdash B \vee A$

$$\dfrac{A \vee B \qquad \dfrac{\overset{1}{A}}{B \vee A}\,\vee I_2 \qquad \dfrac{\overset{1}{B}}{B \vee A}\,\vee I_1}{B \vee A}\,\vee E{,}1$$

It depends only on the form of the premiss and conclusion which of the two I-rules is applied: In $\vee I_1$, the premiss becomes the left disjunct, in $\vee I_2$, the right disjunct.

b. $(A \supset C)\,\&\,(B \supset C) \vdash A \vee B \supset C$

$$\dfrac{\dfrac{\overset{2}{A \vee B} \qquad \dfrac{\dfrac{(A \supset C)\,\&\,(B \supset C)}{A \supset C}\,\&E_1 \quad \overset{1}{A}}{C}\,\supset E \qquad \dfrac{\dfrac{(A \supset C)\,\&\,(B \supset C)}{B \supset C}\,\&E_2 \quad \overset{1}{B}}{C}\,\supset E}{C}\,\vee E{,}1}{A \vee B \supset C}\,\supset I{,}2$$

In the construction of derivations such as the above, it is best to start from the cases A and B, to work towards some common consequence C, and then to add the major premiss $A \vee B$ and its inference line.

Contrary to what one might expect, the converse to (b) does not use disjunction elimination:

c. $A \vee B \supset C \vdash (A \supset C)\,\&\,(B \supset C)$

$$\dfrac{\dfrac{\dfrac{A \vee B \supset C \quad \dfrac{\overset{1}{A}}{A \vee B}\,\vee I_1}{C}\,\supset E}{A \supset C}\,\supset I{,}1 \qquad \dfrac{\dfrac{A \vee B \supset C \quad \dfrac{\overset{2}{B}}{A \vee B}\,\vee I_2}{C}\,\supset E}{B \supset C}\,\supset I{,}2}{(A \supset C)\,\&\,(B \supset C)}\,\&I$$

By derivations (b) and (c), we conclude that the formulas $A \vee B \supset C$ and $(A \supset C)\,\&\,(B \supset C)$ are logically equivalent. That is hardly clear without some reflection. However, if for C we take the formula $\bot$, we obtain as a special case the equivalence:

$$\neg(A \vee B) \supset\subset \neg A\,\&\,\neg B$$

Now we have a logical law the correctness of which can be immediately felt, contrary to the more general form with an arbitrary consequence C of the disjunction $A \vee B$. The equivalence shows that cases (disjunctions) in the antecedent of an implication are not genuine, because any such formula can be replaced by one that does not have the cases.

The premisses of a logical rule must be in some intuitive sense at least as strong as the conclusion. Clearly $A \& B$ is stronger than A, and A is stronger than $A \vee B$. How could one, then, have any use for the two rules for introducing a disjunction? Why make the conclusion $A \vee B$ if you already knew A? Consider the following argument: We are at the second round of a presidential election, with candidates A. Bell and C. Davis left. Bell is a liberal, Davis a female. The conclusion is that either a liberal or a female is chosen as the next president. The formal derivation is as follows, with the atomic formulas:

B is: *Bell is chosen as president.*
D is: *Davis is chosen as president.*
L is: *A liberal is chosen as president.*
F is: *A female is chosen as president.*

We agree that $B \vee D$ is correct because of the electoral arrangement, and that $B \supset L$ and $D \supset F$ are correct conditional sentences. We want to show:

$$B \vee D,\, B \supset L,\, D \supset F \vdash L \vee F$$

Here is the formal derivation:

$$\cfrac{B \vee D \qquad \cfrac{\cfrac{B \supset L \quad \overset{1}{B}}{L}\,{\supset}E}{L \vee F}\,{\vee}I_1 \qquad \cfrac{\cfrac{D \supset F \quad \overset{1}{D}}{F}\,{\supset}E}{L \vee F}\,{\vee}I_2}{L \vee F}\,{\vee}E,1$$

The disjunction $L \vee F$ was introduced in two subderivations under the added assumptions B and D, respectively, that are closed when the overall conclusion is reached.

Let us look again at example (c). The conjunction $A \& B$ is clearly stronger than the disjunction $A \vee B$, so we can make the antecedent of the assumption stronger by assuming $A \& B \supset C$ instead of $A \vee B \supset C$. It happens that $(A \supset C) \& (B \supset C)$ is no longer derivable. Thinking of it, $A \vee B \supset C$ is stronger than $A \& B \supset C$ because it claims C to follow from a weaker antecedent.

(b) Conversion for disjunction. The addition of disjunction to our language provokes some complications in the definition of a normal derivation and as a consequence also in the proof of the normalization of derivations. An introduction can be separated from the corresponding elimination by a step of disjunction elimination, in a situation known as a **permutation convertibility**. Here is an example of a part of derivation with such a convertibility:

Table 3.8 A permutation convertibility

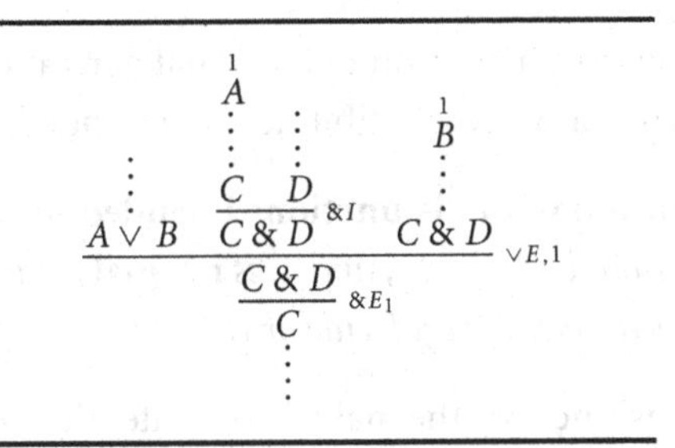

The derivation can be modified so that the last step of the part shown, an instance of rule $\&E$, is permuted to above $\vee E$:

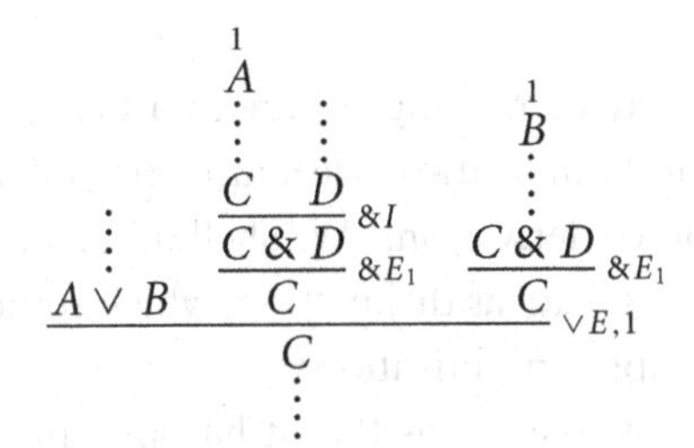

Now there is a detour convertibility on $C \& B$ that can be eliminated as in Section 3.2.

As with the previous connectives, there is also the possibility of a detour convertibility with $\vee$, when a derivation has a part of the form:

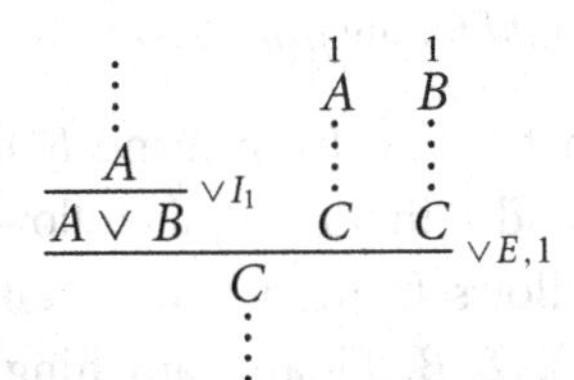

The part is converted by composing parts of the original derivation after the following pattern:

There is an analogous conversion if $A \vee B$ was concluded by rule $\vee I_2$ from B.

With the addition of disjunction, the notion of a normal derivation has to be modified so that also permutative convertibilities are excluded:

Definition 3.2. Normal derivation with disjunctions included. *A derivation in Gentzen's natural deduction for* $\&$, $\vee$, $\supset$, *and* $\perp$ *is* **normal** *if no major premiss of an elimination rule is derived by an I-rule or rule* $\vee E$.

As with the fragment without disjunction, the main result states that derivations convert to normal form.

3.4 A final modification

There is no known way of making disjunction behave in the same neat way as conjunction and implication, with just detour conversions required for the normalization of derivations. We shall, instead, make the rules for the latter connectives as 'bad' as disjunction, which somewhat surprisingly leads to some remarkable simplifications.

One characteristic of rule $\vee E$ is that it has an arbitrary conclusion C, whereas the conclusions of rules $\& E$ and $\supset E$ were components of the major premisses. If we look at the I-rules for disjunction they can be said to give the **sufficient grounds** for introducing a disjunction. One ground is that we have derived A, the other that we have derived B. We now stipulate:

The inversion principle. *Whatever follows from the sufficient grounds for introducing a formula, should follow from that formula.*

With rule $\vee I$, anything that follows from A and B, taken separately because of the two possibilities A and B in $\vee I$, should follow from $A \vee B$. Similarly, for $\& I$, anything that follows from A and B, taken together as in rule $\& I$, should follow from $A \& B$. Finally, anything that follows from the derivability of B from the assumption A, should follow from $A \supset B$. The way to look at the derivability of B from A is that it can be used for reducing arbitrary consequences of B, call them C, into consequences of A. We have thus the two elimination rules that we call **general**:

Table 3.9 General elimination rules

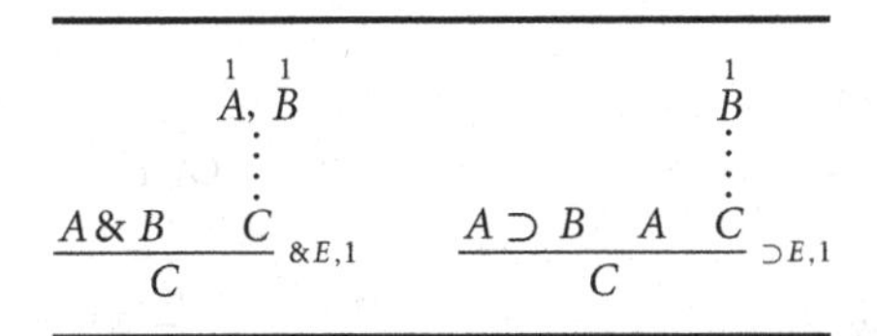

In rule $\&E$, we have first that C is a consequence of A and B together. By the rule, it is then a consequence of $A \& B$. Similarly, C is a consequence of B and becomes by rule $\supset E$ a consequence of $A \supset B$ and A.

Gentzen's E-rules are special cases of the general elimination rules: If in the first rule of Table 3.9 we set $C \equiv A$ and then $C \equiv B$, and $C \equiv B$ in the second rule, we get:

Table 3.10 Gentzen's rules as special cases of general rules

$$\frac{A \& B \quad \overset{1}{A}}{A}\,{\scriptstyle \&E,1} \qquad \frac{A \& B \quad \overset{1}{B}}{B}\,{\scriptstyle \&E,1} \qquad \frac{A \supset B \quad A \quad \overset{1}{B}}{B}\,{\scriptstyle \supset E,1}$$

The derivations of the rightmost premiss are degenerate in each case and can be left unwritten which gives exactly the Gentzen rules. In the other direction, the conclusions of the general rules can be derived from their premisses by Gentzen's rules: Assume there to be derivations of C from A, B as in the premiss of rule $\&E$ of Table 3.9, and of C from B as in the premiss of rule $\supset E$ in the same table. Then the conclusions of the general rules are obtained from their premisses by Gentzen's rules and compositions:

$$\begin{array}{c} \dfrac{A \& B}{A}\,{\scriptstyle \&E_1} \quad \dfrac{A \& B}{B}\,{\scriptstyle \&E_2} \\ \vdots \\ C \end{array} \qquad\qquad \begin{array}{c} \dfrac{A \supset B \quad A}{B}\,{\scriptstyle \supset E} \\ \vdots \\ C \end{array}$$

The modified system with the general E-rules is equivalent to Gentzen's system as far as the derivability of formulas from given assumptions is concerned.

If the conclusion C of an elimination rule is the major premiss of another elimination rule, the latter can now be permuted up in all the rules, not just for $\vee E$ as with Gentzen's rules. There are three E-rules (plus $\bot E$) so that there are altogether nine possible combinations, $\&E$ and $\&E$, $\&E$ and $\vee E$, etc. Here is the first one together with the converted derivation part:

$$\dfrac{\dfrac{\begin{matrix}\vdots\\ A\&B\end{matrix} \quad \begin{matrix}\overset{1}{A},\ \overset{1}{B}\\ \vdots\\ C\&D\end{matrix}}{C\&D}\,{\scriptstyle \&E,1} \quad \begin{matrix}\overset{2}{C},\ \overset{2}{D}\\ \vdots\\ E\end{matrix}}{E}\,{\scriptstyle \&E,2} \qquad\qquad \dfrac{\begin{matrix}\vdots\\ A\&B\end{matrix} \quad \dfrac{\begin{matrix}\overset{1}{A},\ \overset{1}{B}\\ \vdots\\ C\&D\end{matrix} \quad \begin{matrix}\overset{2}{C},\ \overset{2}{D}\\ \vdots\\ E\end{matrix}}{E}\,{\scriptstyle \&E,2}}{E}\,{\scriptstyle \&E,1}$$

With $\bot E$, there is the possibility of its repeated application that is eliminated as in the transformation:

$$\dfrac{\dfrac{\bot}{\bot}\,{\scriptstyle \bot E}}{C}\,{\scriptstyle \bot E} \qquad\qquad \dfrac{\bot}{C}\,{\scriptstyle \bot E}$$

We can give our final definition of a normal derivation, this time without the exception on $\vee E$ of Gentzen's definition:

Definition 3.3. Normal derivation with general elimination rules. *A derivation with general elimination rules is* **normal** *if all major premisses of E-rules are assumptions.*

As an example of the general form of normal derivations, consider the following example:

$$\dfrac{\dfrac{(A\&B)\&C}{A\&B}\,{\scriptstyle \&E}}{A}\,{\scriptstyle \&E}$$

This derivation of A from the assumption $(A\&B)\&C$ is normal in Gentzen's sense. It has, however, a major premiss of an E-rule that has been derived, namely $A\&B$. The derivation, considered as a special case of a derivation with the general rules, is:

$$\dfrac{\dfrac{(A\&B)\&C \quad \overset{1}{A\&B}}{A\&B}\,{\scriptstyle \&E,1} \quad \overset{2}{A}}{A}\,{\scriptstyle \&E,2}$$

Here we see that the lower instance of rule $\&E$ is not normal in the general sense. A permutative conversion moves the lower instance up, with a normal derivation as a result:

$$\dfrac{(A\,\&\,B)\,\&\,C \qquad \dfrac{\overset{1}{A\,\&\,B} \quad \overset{2}{A}}{A}{\scriptstyle \&E,2}}{A}{\scriptstyle \&E,1}$$

The most important property of normal derivations is:

Subformula property. *All formulas in a normal derivation are subformulas of open assumptions or of the endformula of the derivation.*

We shall see in the next chapter how the decidability of derivability follows from the subformula property.

Assume given a normal derivation of a theorem, i.e., a derivation that has no open assumptions left. Were the last rule an elimination, the major premiss would be an assumption that has not been closed. Therefore we have:

Property of direct provability. *Proofs of theorems convert to a form in which the last rule is an introduction.*

An immediate consequence is that if $A \vee B$ is a theorem, the last rule in a normal derivation must be $\vee I$. Therefore one of A or B is a theorem:

Disjunction property. *If $A \vee B$ is derivable, one of A and B is derivable.*

The result fails, naturally, if $A \vee B$ has been derived from assumptions that contain essential disjunctions.

A proof of the normalizability of derivations is found in Section 13.2. Dag Prawitz and Andres Raggio had given proofs of normalization that were published, independently of each other, in 1965. We found a proof of normalization by Gentzen in a handwritten early version of his thesis in February 2005, published in English translation in 2008. Gentzen's proof was completed perhaps during the first months of 1933, but for reasons that will be explained later, he had no use for it in its original form.

Gentzen's proof proceeds as follows: First any possible permutation convertibilities are located. As in Table 3.8, there is a second occurrence of the major premiss of the last elimination ($C\,\&\,D$ in Table 3.8), namely above the inference line for $\vee E$. A derivation can be such that this succession of the same formula repeats itself over and over again. However, after a permutative conversion, there is just one occurrence of $C \& D$ in the example of Table 3.8. In general, a permutative conversion diminishes the occurrence of the formula in succession by one, until there are no permutation convertibilities left. Then detour conversions are applied. These can lead to new convertibilities, but they are on shorter formulas. In the end, after a bounded number of conversions, a normal derivation is reached.

3.5 The role of falsity and negation

(a) The rule of falsity elimination. We have not paid much attention to falsity and its rule $\bot E$. There is a familiar rule of logic that has the same deductive strength as rule $\bot E$. Consider the following derivation:

$$\dfrac{A \lor B \quad \overset{1}{A} \quad \dfrac{\dfrac{\neg B \quad \overset{1}{B}}{\bot}{\scriptstyle \supset E}}{A}{\scriptstyle \bot E}}{A}{\scriptstyle \lor E,1}$$

The derivation shows that from $A \lor B$ and $\neg B$, the conclusion A can be drawn. We could take it as a rule that is added to our system, called by its mediæval name *modus tollendo ponens* (something like 'mode of positing by taking away'):

$$\dfrac{A \lor B \quad \neg B}{A}{\scriptstyle\, mtp}$$

Logicians have studied what happens if rule $\bot E$ is left out of the system of natural deduction. What remains, i.e., the introduction and elimination rules for the connectives &, $\lor$, and $\supset$, is called **minimal** propositional logic. It is known that the conclusion of rule *mtp* is not derivable from its premisses if rule $\bot E$ is left out. Rule *mtp* has in fact the same deductive strength as rule $\bot E$: The above showed that the conclusion of *mtp* is derivable from its premisses by rules $\supset E$, $\bot E$, and $\lor E$. In the other direction, assume there to be a derivation of $\bot$ from some given assumptions. The task is to show that the conclusion of $\bot E$, i.e., any formula C, is derivable by *mtp*:

$$\dfrac{\dfrac{\begin{matrix}\vdots\\ \bot\end{matrix}}{C \lor \bot}{\scriptstyle \lor I_2} \quad \dfrac{\overset{1}{\bot}}{\neg \bot}{\scriptstyle \supset I,1}}{C}{\scriptstyle\, mtp}$$

Rule $\supset I$ is used to derive $\neg\bot$, i.e., $\bot \supset \bot$. There might be sometimes specific formal reasons for not using rule $\bot E$, but other possible objections to the rule, if any, should apply as well to rule *mtp*.

Finally, we note that even in minimal logic, any negative formula can be inferred from $\bot$:

$$\frac{\bot}{\neg C}\supset I$$

This is an instance of rule $\supset I$ in which the assumption C was in fact not used.

(b) Derivable and admissible rules. By the derivation that began this section, *mtp* is a **derivable rule** in natural deduction. In general:

Definition 3.4. Derivable rules. *A rule is* **derivable** *in a given system of rules if its conclusion is derivable from its premisses in the system.*

There is a weaker but in fact more important notion, called **admissibility**:

Definition 3.5. Admissible rules. *A rule is* **admissible** *in a given system of rules if its conclusion is derivable in the system whenever its premisses are.*

A derivable rule is admissible, but not necessarily the other way around.

Proofs of admissibility proceed by considering the way or ways in which the premisses can have been derived. A good example of an admissible rule in natural deduction is the composition of two derivations, as in Table 3.4: Whenever there is a derivation of A from the assumptions Γ and a derivation of C from the assumptions A, Δ, it is permitted to conclude C from the assumptions Γ, Δ.

It is not unusual to mix the notions of admissibility and derivability, even in textbooks on logic. The argument goes like this:

Assume that if A is derivable, also B is. Then $A \supset B$ *is derivable.* This step is a plain **logical fallacy** to which we shall return in Section 3.7.

(c) Negation as a primitive notion. It is possible to use a primitive notion of negation, instead of the one defined through implication and the special formula $\bot$. The rules of primitive negation are two, an introduction and an elimination:

Table 3.11 Rules for primitive negation

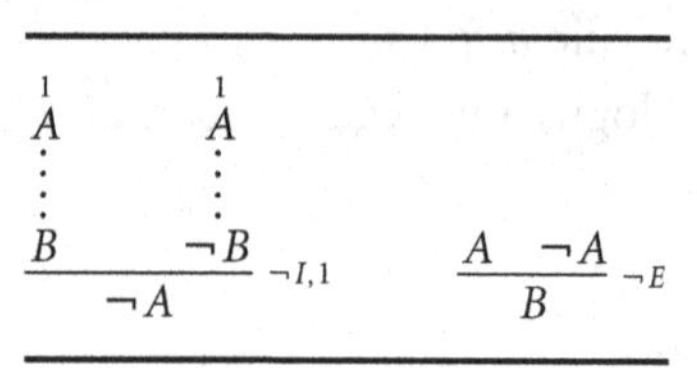

Assumption A would in many cases appear only above one of the premisses of the I-rule.

The negation rules are equivalent to ones with negation defined through $\bot$. We show first that the conclusions of the new rules follow from their premisses by the earlier rules: For rule $\neg I$, assume there to be derivations of its premisses, then apply $\supset E$ to conclude $\bot$, next $\supset I$ to conclude $\neg A$. For the second rule, apply $\supset E$ to conclude $\bot$, next $\bot E$ to conclude B.

The other direction goes so that occurrences of $\bot$ in derivations are replaced by contradictions of the form $P \,\&\, \neg P$: If A is an assumption with a part of the form $\supset \bot$, replace $\bot$ by $P \,\&\, \neg P$ with P, say, the first atom of A. Whenever a rule other than one of $\supset I$, $\supset E$, or $\bot E$ is applied, nothing more need be done. If the last step in a derivation is $\supset I$, the derivation and its transformation are:

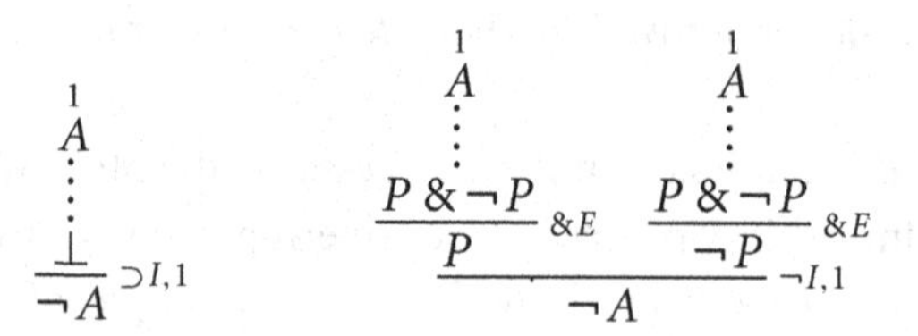

Here P can be chosen from A as before. The transformations for the other remaining rules are similar.

In Gentzen's thesis, the rules of natural deduction are given so that negation can be taken either as defined or as primitive, but later the defined notion is used. The latter choice gives a greater uniformity to the transformation of derivations into normal form. With primitive negation, we have the possibility of repeated instances of $\neg I$, and of $\neg I$ followed by $\neg E$:

$$\dfrac{\overset{2}{C} \;\vdots\; A \qquad \dfrac{\overset{1}{A},\overset{2}{C} \;\vdots\; B \qquad \overset{1}{A},\overset{2}{C} \;\vdots\; \neg B}{\neg A}{\scriptstyle \neg I,1}}{\neg C}{\scriptstyle \neg I,2} \qquad\qquad \dfrac{\vdots\; A \qquad \dfrac{\overset{1}{A} \;\vdots\; B \qquad \overset{1}{A} \;\vdots\; \neg B}{\neg A}{\scriptstyle \neg I,1}}{C}{\scriptstyle \neg E}$$

These two situations can be converted into forms that are obviously simpler:

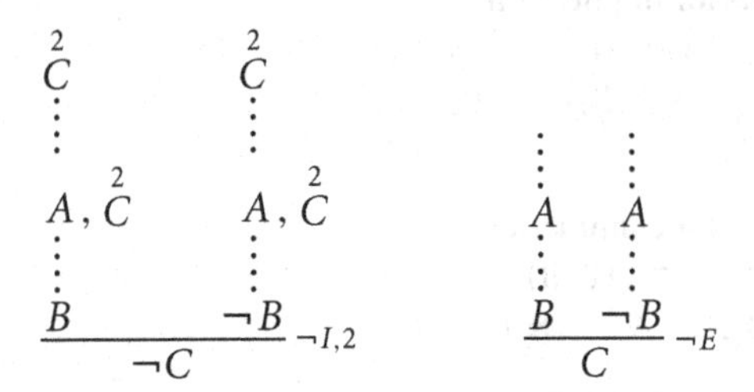

One negation and one rule instance has disappeared from each derivation, but there is no further systematic sense in which these conversions would give simpler derivations as results. If the unconverted derivations are translated into the standard calculus with a defined negation, it is instead seen that both of the successions of rules $\neg I, \neg I$ and $\neg I, \neg E$ contain a detour convertibility. Moreover, the simplified derivations correspond exactly to the results of detour conversions. The details are left as an exercise.

3.6 Axiomatic logic

Before Gentzen's rule-based proof systems were developed in the early 1930s, logic had a fifty-year **axiomatic** tradition. It began with Frege, as explained in Section 14.3, and paralleled similar developments in mathematics. One of the main impulses here was the invention of non-Euclidean geometries, and especially the question of the independence of the parallel postulate. The question of what can be proved is meaningful only if the principles of proof have been explicitly laid down.

(a) The axiomatization of logic. By the 1920s, logic was axiomatized in the style of geometry, so that each of its basic notions had a separate group of axioms. Different but equivalent systems of axiomatization were proposed, each of them inevitably mixing a bit the basic notions, for example, each axiom requires an implication or a negation. A pure formulation with no such mixing became possible only with the invention of natural deduction. The following collection of axioms is similar to that of Hilbert and Bernays (cf. Section 14.4), except for the negation axioms.

Table 3.12 Axiomatic logic

I Axioms for implication

1. $A \supset (B \supset A)$
2. $(A \supset (A \supset B)) \supset (A \supset B)$
3. $(A \supset B) \supset ((B \supset C) \supset (A \supset C))$

II Axioms for conjunction

4. $A \supset (B \supset A \& B)$
5. $A \& B \supset A, \quad A \& B \supset B$

III Axioms for disjunction

6. $A \supset A \vee B, \quad B \supset A \vee B$
7. $(A \supset C) \supset ((B \supset C) \supset (A \vee B \supset C))$

IV Axioms for negation

8. $(A \supset B) \supset (A \supset \neg B) \supset \neg A$
9. $A \supset (\neg A \supset C)$

The only rule of inference is $\supset E$. Any instances of the axioms as well as previously proved theorems can be used as premisses in the rule.

Derivations in axiomatic logic are notoriously hard to find. All rules are instances of $\supset E$ so that from an endformula C, the theorem proved, a leftmost branch passes always through major premisses to a topformula that is an axiom instance. It has the form $A_1 \supset (A_2 \supset \ldots \supset (A_n \supset C)\ldots)$. The minor premisses are, from the root, $A_n, A_{n-1}, \ldots, A_1$. Each of them has the same structure, a leftmost branch that ends with an axiom instance. Derivations consist in chopping off from axioms, all of them implications, the antecedents until the endformula is reached. There is no way of transforming a derivation, but it is a completely static object.

What are the advantages of putting on proofs such a one-rule straitjacket as in axiomatic logic, with the ensuing enormous difficulty of finding the right axiom instances? One reason may be historical: In the axiomatic studies of geometry from the latter part of the nineteenth century on, there was a tendency to minimize the number of basic concepts and the number of axioms about them. Another reason is that it is sometimes very easy to prove properties of a logical calculus if it is put in the axiomatic form. If all the axioms have some property of interest, if rule $\supset E$ maintains that property, then also all theorems have the property.

We have met most of the above axioms as examples of theorems in intuitionistic logic, and also the remaining ones are easily proved. Therefore we can define:

Translation from axiomatic logic to natural deduction. *Given a derivation in axiomatic logic, it is translated into natural deduction by substituting the instances of axioms in the derivation by the derivation of these axioms in natural deduction.*

The result of such a translation produces a great number of non-normalities: Each axiom is an implication, often put into use as the major premiss of rule $\supset E$. A normal derivation of the axiom has rule $\supset I$ as a last rule, so that a detour convertibility is produced.

The reverse translation from natural deduction to axiomatic logic is a lot trickier. First of all, there is no notion of derivation from assumptions in axiomatic logic as it was originally conceived. Frege and Russell took the axioms to express logical truths, and each step of proof just produced a new logical truth B from previous truths $A \supset B$ and A. Later, when axiomatic logic was applied to abstract axiom systems, this idea had to be changed. The collection of assumptions in the following definition is a list of formulas in which the order plays no role:

Definition 3.6. Derivations from assumptions in axiomatic logic. *A formula C is derivable from a collection of assumptions Γ in axiomatic logic in the following cases:*

1. *C is an instance of an axiom.*
2. *C is one of the assumptions Γ.*
3. *There are derivations of $A \supset C$ from the assumptions Δ and of A from the assumptions Θ and Γ is the collection Δ, Θ.*

We shall write rule $\supset E$ in derivations in axiomatic logic with assumptions in the form:

$$\frac{\Delta \vdash A \supset C \quad \Theta \vdash A}{\Delta, \Theta \vdash C} {\scriptstyle \supset E}$$

The assumptions are put together in the conclusion of the rule, into a collection Δ, Θ.

If a premiss comes from clause 1 of the above definition, it has the form $\Gamma \vdash C$ in which the collection of assumptions Γ can be chosen freely, or left empty if needed. Similarly for clause 2, it can be written as $\Gamma \vdash C$ with just the condition that C be one of the formulas in Γ. It is essential that all assumptions are counted in Γ, also multiple occurrences.

(b) The deduction theorem. In the translation of derivations in natural deduction to axiomatic logic, a pattern repeats itself in which rule $\supset I$ is applied in natural deduction. Its image under the translation to be defined is known as 'the deduction theorem':

Lemma 3.7. The deduction theorem for axiomatic logic. *If* $A, \Gamma \vdash C$, *then* $\Gamma \vdash A \supset C$.

Proof. We go through the three cases of definition 3.6 and show that the deduction theorem applies in the base cases 1 and 2, and that its applicability is maintained under rule $\supset E$:

1. Formula C is an instance of an axiom. We take an instance of axiom I.1 in Table 3.12 as a first premiss, choose Γ as the collection of assumptions in a second premiss, and have the derivation:

$$\frac{\vdash C \supset (A \supset C) \quad \Gamma \vdash C}{\Gamma \vdash A \supset C}\supset E$$

2. Formula C is one of the assumptions. If it is in Γ, the same derivation as in 1 gives the result. If $C \equiv A$, the derivation is, with instances of axioms I.2 and I.1, respectively:

$$\frac{\vdash (A \supset (A \supset A)) \supset (A \supset A) \quad \Gamma \vdash A \supset (A \supset A)}{\Gamma \vdash A \supset A}\supset E$$

We have now shown that with the starting points of derivations in axiomatic logic with assumptions, namely clauses 1 and 2, a formula can be lifted from the assumptions to the other side of the turnstile, into the antecedent of an implication. The final task is to show that the rule of inference maintains this property, i.e., that whenever its premisses obey the deduction theorem, also the conclusion does:

3. $A, \Gamma \vdash C$ has been derived by rule $\supset E$ from two premisses, either $A, \Gamma' \vdash B \supset C$ and $\Gamma'' \vdash B$, or $\Gamma' \vdash B \supset C$ and $A, \Gamma'' \vdash B$, with $\Gamma \equiv \Gamma', \Gamma''$. The first case is:

$$\frac{\begin{matrix}\vdots \\ A, \Gamma' \vdash B \supset C\end{matrix} \quad \begin{matrix}\vdots \\ \Gamma'' \vdash B\end{matrix}}{A, \Gamma', \Gamma'' \vdash C}\supset E$$

Application of the deduction theorem to the first premiss of rule $\supset E$ gives $\Gamma' \vdash A \supset (B \supset C)$. The next step is to derive the formula:

$$(A \supset (B \supset C)) \supset (B \supset (A \supset C))$$

This is left as an exercise. Then we have the derivation, in which DT indicates an application of the deduction theorem to the premiss $A, \Gamma' \vdash B \supset C$:

$$\dfrac{\dfrac{\vdash (A \supset (B \supset C)) \supset (B \supset (A \supset C)) \qquad \dfrac{\begin{array}{c}\vdots \\ A, \Gamma' \vdash B \supset C\end{array}}{\Gamma' \vdash A \supset (B \supset C)}{\scriptstyle DT}}{\Gamma' \vdash B \supset (A \supset C)}{\scriptstyle \supset E} \qquad \begin{array}{c}\vdots \\ \Gamma'' \vdash B\end{array}}{\Gamma', \Gamma'' \vdash A \supset C}{\scriptstyle \supset E}$$

The second case is that A is in the second premiss $A, \Gamma'' \vdash B$. The derivation is:

$$\dfrac{\begin{array}{c}\vdots \\ \Gamma' \vdash B \supset C\end{array} \quad \begin{array}{c}\vdots \\ A, \Gamma'' \vdash B\end{array}}{A, \Gamma', \Gamma'' \vdash C}{\scriptstyle \supset E}$$

Analogously to the first case, we apply DT to the second premiss, and then use an instance of axiom I.3:

$$\dfrac{\dfrac{\vdash (A \supset B) \supset ((B \supset C) \supset (A \supset C)) \qquad \dfrac{\begin{array}{c}\vdots \\ A, \Gamma'' \vdash B\end{array}}{\Gamma'' \vdash A \supset B}{\scriptstyle DT}}{\Gamma'' \vdash (B \supset C) \supset (A \supset C)}{\scriptstyle \supset E} \qquad \begin{array}{c}\vdots \\ \Gamma' \vdash B \supset C\end{array}}{\Gamma', \Gamma'' \vdash A \supset C}{\scriptstyle \supset E}$$

QED.

The deduction theorem in the above form will take care of the translation of such implication introductions in which exactly one occurrence of an open assumption is closed. Our next task is to deal with vacuous and multiple discharge.

Lemma 3.8. Vacuous deduction theorem. *If* $\Gamma \vdash C$, *then even* $\Gamma \vdash A \supset C$.

Proof. The following derivation gives a proof, with the major premiss an instance of axiom I.1:

$$\dfrac{\vdash C \supset (A \supset C) \quad \Gamma \vdash C}{\Gamma \vdash A \supset C}{\scriptstyle \supset E}$$

QED.

Lemma 3.9. Multiple deduction theorem. *If* $A^m, \Gamma \vdash C$, *then* $\Gamma \vdash A \supset C$.

Proof. We assume first that $m = 2$ and show how to move the two copies of formula A from the antecedent into one in the succedent. Note that by lemma 3.7, repeated application of DT to the premiss $A, A, \Gamma \vdash C$ is justified. The following derivation gives a proof, with the major premiss an instance of axiom I.2:

$$\dfrac{\vdash (A \supset (A \supset C)) \supset (A \supset C) \quad \dfrac{\dfrac{A, A, \Gamma \vdash C}{A, \Gamma \vdash A \supset C}{\scriptstyle DT}}{\Gamma \vdash A \supset (A \supset C)}{\scriptstyle DT}}{\Gamma \vdash A \supset C}{\scriptstyle \supset E}$$

With more than two copies of *A*, the procedure is repeated. QED.

Applications of the deduction theorem can be represented as instances of rule *DT*:

$$\dfrac{A, \Delta \vdash C}{\Delta \vdash A \supset C}{\scriptstyle DT}$$

The proof of the deduction theorem shows how instances of rule *DT* can be eliminated from derivations in axiomatic logic. It is therefore an admissible rule in axiomatic logic with assumptions, a notion defined in Section 3.5(b).

The deduction theorem has a couple of useful corollaries, the first of which tells us that derivations in axiomatic logic with assumptions are **closed under composition.**

Corollary 3.10. Closure under composition in axiomatic logic. *If* $\Gamma \vdash A$ *and* $A, \Delta \vdash C$, *then* $\Gamma, \Delta \vdash C$.

Proof. By the derivation

$$\dfrac{\dfrac{A, \Delta \vdash C}{\Delta \vdash A \supset C}{\scriptstyle DT} \quad \Gamma \vdash A}{\Gamma, \Delta \vdash C}{\scriptstyle \supset E}$$

QED.

As with the deduction theorem, composition can be given as an admissible rule:

$$\dfrac{\Gamma \vdash A \quad A, \Delta \vdash C}{\Gamma, \Delta \vdash C}{\scriptstyle Comp}$$

A further corollary shows that rule *DT* is invertible in the sense of the following 'inverse rule of detachment':

Corollary 3.11. Inverse deduction theorem. *If* $\Gamma \vdash A \supset C$, *then* $A, \Gamma \vdash C$.

Proof. By the derivation

$$\dfrac{\Gamma \vdash A \supset C \quad A \vdash A}{A, \Gamma \vdash C}{\scriptstyle \supset E}$$

QED.

An example will show how the composition of derivations works in axiomatic logic. Here are two derivations that are composed, with the **composition formula** $A \,\&\, B$. To make the compositions fit a page, we write conjunctions in the style of AB:

$$\dfrac{\dfrac{\vdash (AB)C \supset AB \quad (AB)C \vdash (AB)C}{(AB)C \vdash AB}\supset E \qquad \dfrac{\vdash (AB) \supset A \quad AB \vdash AB}{AB \vdash A}\supset E}{(AB)C \vdash A}\,Comp$$

Rule *Comp* is eliminated as in the proof of corollary 3.10:

$$\dfrac{\dfrac{\dfrac{\vdash AB \supset A \quad AB \vdash AB}{AB \vdash A}\supset E}{\vdash AB \supset A}\,DT \qquad \dfrac{\vdash (AB)C \supset AB \quad (AB)C \vdash (AB)C}{(AB)C \vdash AB}\supset E}{(AB)C \vdash A}\supset E$$

The final step is to eliminate DT. Looking at the proof of the deduction theorem, we notice that the formula that is moved in DT comes from the right premiss in the instance of rule $\supset E$ that is used to derive the premiss of DT. Therefore the procedure of elimination is as in the second case of 3 in the proof of lemma 3.7. The deduction theorem is applied to the second premiss, followed by an instance of axiom I.3 and two instances of $\supset E$. We cannot write the whole of the above derivation thus transformed, because it is too broad, but write just the subderivation of the left premiss, the one concluded by *DT:*

$$\dfrac{\dfrac{\vdash (AB \supset AB) \supset ((AB \supset A) \supset (AB \supset A)) \quad \dfrac{AB \vdash AB}{\vdash AB \supset AB}\,DT}{\vdash (AB \supset A) \supset (AB \supset A)}\supset E \qquad \vdash AB \supset A}{\vdash AB \supset A}\supset E$$

The remaining instance of DT is eliminated as in the proof of the deduction theorem, case 2, but the derivation becomes again too broad.

The result of the above transformation has lots of redundancies in the form of loops. When these are eliminated, a printable derivation is obtained:

$$\dfrac{\vdash AB \supset A \qquad \dfrac{\vdash (AB)C \supset AB \quad (AB)C \vdash (AB)C}{(AB)C \vdash AB}\supset E}{(AB)C \vdash A}\supset E$$

The next question is to translate derivations from axiomatic logic with assumptions into derivations without assumptions, and to show that if a formula C is derivable in the former from no assumptions, it is derivable also in the latter. We leave these tasks as advanced exercises and turn instead to the next topic.

(c) Translation from natural deduction to axiomatic logic. A derivation in natural deduction is translated into axiomatic logic in two stages: First each formula C in the derivation is replaced by $\Gamma \vdash C$ in which Γ is the collection of open assumptions, multiplicity counted, on which the occurrence of C depends. In a second stage, the result is transformed rule by rule:

Conjunction rules. With $\&I$, we have after stage 1:

$$\dfrac{\begin{array}{cc}\vdots & \vdots \\ \Gamma \vdash A & \Delta \vdash B\end{array}}{\begin{array}{c}\Gamma, \Delta \vdash A \& B \\ \vdots\end{array}}\,\&I$$

Axiom II.4 is used in stage 2:

$$\dfrac{\dfrac{\vdash A \supset (B \supset A \& B) \quad \begin{array}{c}\vdots \\ \Gamma \vdash A\end{array}}{\Gamma \vdash B \supset A \& B}\supset E \qquad \begin{array}{c}\vdots \\ \Delta \vdash B\end{array}}{\begin{array}{c}\Gamma, \Delta \vdash A \& B \\ \vdots\end{array}}\supset E$$

Rule $\&E$ is translated similarly, with the two stages and axiom II.5 in the latter:

$$\dfrac{\begin{array}{c}\vdots \\ \Gamma \vdash A \& B\end{array}}{\begin{array}{c}\Gamma \vdash A \\ \vdots\end{array}}\,\&E \qquad \dfrac{\vdash A \& B \supset A \quad \begin{array}{c}\vdots \\ \Gamma \vdash A \& B\end{array}}{\begin{array}{c}\Gamma \vdash A \\ \vdots\end{array}}\supset E$$

The second elimination rule is entirely similar.

Disjunction rules. Rule $\vee I$ is translated in the two stages, with axiom III.6 in the latter:

$$\dfrac{\begin{array}{c}\vdots \\ \Gamma \vdash A\end{array}}{\begin{array}{c}\Gamma \vdash A \vee B \\ \vdots\end{array}}\,\vee I \qquad \dfrac{\vdash A \supset A \vee B \quad \begin{array}{c}\vdots \\ \Gamma \vdash A\end{array}}{\begin{array}{c}\Gamma \vdash A \vee B \\ \vdots\end{array}}\supset E$$

The second introduction rule is entirely similar.

With $\vee E$, we have after stage 1:

$$\dfrac{\begin{array}{ccc}\vdots & \vdots & \vdots \\ \Gamma \vdash A \vee B & A^m, \Delta \vdash C & B^n, \Theta \vdash C\end{array}}{\begin{array}{c}\Gamma, \Delta, \Theta \vdash C \\ \vdots\end{array}}\,\vee E$$

The minor premisses are first transformed by the deduction theorem, and then axiom III.7 is used:

$$\dfrac{\dfrac{\dfrac{\vdash (A \supset C) \supset ((B \supset C) \supset (A \lor B \supset C)) \quad \dfrac{\overset{\vdots}{A^m, \Delta \vdash C}}{\Delta \vdash A \supset C}\,DT}{\Delta \vdash (B \supset C) \supset (A \lor B \supset C)}\supset E \quad \dfrac{\overset{\vdots}{B^n, \Theta \vdash C}}{\Theta \vdash B \supset C}\,DT}{\Delta, \Theta \vdash A \lor B \supset C}\supset E \quad \overset{\vdots}{\Gamma \vdash A \lor B}}{\underset{\vdots}{\Gamma, \Delta, \Theta \vdash C}}\supset E$$

Implication rules. Rule $\supset I$ is covered by the deduction theorem and rule $\supset E$ is maintained as it is.

Negation rules. We shall translate the rules of natural deduction with a primitive negation, as in Table 3.11, because that is the way negation is usually treated in axiomatic logic.

For rule $\neg I$, the first stage gives:

$$\dfrac{\overset{\vdots}{A^m, \Gamma \vdash B} \quad \overset{\vdots}{A^n, \Delta \vdash \neg B}}{\underset{\vdots}{\Gamma, \Delta \vdash \neg A}}\,\neg I$$

We apply the deduction theorem to the premisses together with an instance of axiom IV.8:

$$\dfrac{\dfrac{\vdash (A \supset B) \supset ((A \supset \neg B) \supset \neg A) \quad \dfrac{\overset{\vdots}{A^m, \Gamma \vdash B}}{\Gamma \vdash A \supset B}\,DT}{\Gamma \vdash (A \supset \neg B) \supset \neg A}\supset E \quad \dfrac{\overset{\vdots}{A^n, \Delta \vdash \neg B}}{\Delta \vdash A \supset \neg B}\,DT}{\underset{\vdots}{\Gamma, \Delta \vdash \neg A}}\supset E$$

For rule $\neg E$, we have the two stages and axiom IV.9 in the latter:

$$\dfrac{\overset{\vdots}{\Gamma \vdash A} \quad \overset{\vdots}{\Delta \vdash \neg A}}{\underset{\vdots}{\Gamma, \Delta \vdash C}}\,\neg E \qquad \dfrac{\dfrac{\vdash A \supset (\neg A \supset C) \quad \overset{\vdots}{\Gamma \vdash A}}{\Gamma \vdash \neg A \supset C}\supset E \quad \overset{\vdots}{\Delta \vdash \neg A}}{\underset{\vdots}{\Gamma, \Delta \vdash C}}\supset E$$

The axioms in the beginning of this section were chosen so that the translations went through smoothly.

3.7 Proofs of unprovability

The structure of normal derivations makes it possible to show that some formulas are not derivable in our system of natural deduction. For example, a contradiction is not derivable, the law of excluded middle, $A \vee \neg A$, is not derivable, and no equivalence that would show one connective to be definable by the others is derivable.

(a) **Consistency and excluded middle.** By the normalizability of derivations in natural deduction, $\bot$ is not derivable. If it were, it would have a normal derivation and normal derivations must end with an introduction. By the subformula property, there would appear a connective in the endformula, the one of the last I-rule, which is not the case. Therefore the system is of natural deduction **consistent**.

If a **contradiction** $A \& \neg A$ were derivable, both $\neg A$ and A would be derivable by rule $\& E$, so we would have by rule $\supset E$ a derivation of $\bot$, against the consistency of the system. Therefore $A \& \neg A$ is not derivable. On the other hand, if $\bot$ were derivable, $A \& \neg A$ would be derivable by rule $\bot E$. Thus, consistency in the sense of underivability of $\bot$ and freedom of contradiction in the sense of underivability of $A \& \neg A$ are equivalent properties.

Consider next what is known as the **law of excluded middle** that is the characteristic mark of **classical propositional logic**: $A \vee \neg A$. Assume it to be derivable. This means, as explained in Section 2.6, that there is a schematic derivation with $A \vee \neg A$ as the endformula. Substitute for A some atomic formula P and you get a derivation of $P \vee \neg P$. By the disjunction property, P or $\neg P$ is derivable. No I-rule can have the former as a conclusion, so it is not derivable. Therefore $\neg P$ must be derivable. The last rule in a normal derivation is $\supset I$ so that $\bot$ must be derivable from P. The formulas that can appear in a normal derivation are P and $\bot$, but no E-rule applies to P, and I-rules would introduce a connective. Therefore the law of excluded middle is not derivable.

Consider next the **law of double negation**: $\neg\neg A \supset A$. If it is derivable, the instance with $A \vee \neg A$ in place of A, i.e., $\neg\neg(A \vee \neg A) \supset A \vee \neg A$, is in particular derivable. By example (b) in Section 4.1 below,

$\neg\neg(A \vee \neg A)$ is derivable, so that $A \vee \neg A$ is derivable. This cannot be, because we just showed that the law of excluded middle is not derivable. Therefore the law of double negation is not derivable.

The above argument showed that if $\neg\neg A \supset A$ is derivable in intuitionistic logic, also $A \vee \neg A$ is. Thus, that step constitutes an admissible rule. As mentioned at the end of Section 3.5(b), it is a logical fallacy to conclude from such admissibility that the implication $(\neg\neg A \supset A) \supset (A \vee \neg A)$ is derivable. The next chapter gives a systematic means for demonstrating such fallacies, in the form of **underivability results.**

(b) Independence of the connectives. We have seen that there are connections between the connectives, such as the derivable implications $A \,\&\, B \supset \neg(\neg A \vee \neg B)$, $A \vee B \supset \neg(\neg A \,\&\, \neg B)$, and $(A \supset B) \supset \neg(A \,\&\, \neg B)$. It can be shown that none of the converse implications is derivable in intuitionistic logic. Further similar results show the following:

Independence of the connectives in intuitionistic logic. *None of the connectives can be expressed equivalently by the remaining ones.*

It may seem strange that, say, $A \vee B$ is not equivalent to $\neg(\neg A \,\&\, \neg B)$. We shall gain a deeper insight into this matter towards the end of the next chapter.

(c) The point of intuitionistic logic. Why is $\neg\neg A$ not equivalent to A? The former can be read as: *$\neg A$ will eventually turn out impossible.* We have here a **negative** or **indirect** claim, not a positive, direct one. A specific example could be: Take some property of natural numbers, designated $P(n)$ and read as *natural number n has the property P.* Consider the claim *there is a natural number with the property P.* A direct argument for this claim consists in showing a specific natural number and a proof that it has the property P. An indirect argument, instead, would run as follows: Assume for the sake of the argument the contrary, *there is no natural number with the property P,* and let this assumption turn out impossible. Thus, *it is not the case that there is no natural number with the property P.* The proof of this double negation can be such that it in no way shows what number could have the property P. If *it is not the case that there is no natural number with the property P,* we can start running though the natural numbers, $0, 1, 2, \ldots$ and check for the property P. It is impossible that a number with the property would not eventually turn up, but this knowledge need not help in determining how far we have to proceed. Herein lies the difference between the direct statement and its double negation. The phenomenon is general, not tied to examples such as the sequence of natural numbers.

Gentzen's system of natural deduction, and its slight modification as in Section 3.4, gives a system of intuitionistic propositional logic. As mentioned, the term comes from the philosophy of mathematics of L. Brouwer and goes back to the philosophy of Immanuel Kant in the eighteenth century. Brouwer was the first one, exactly one hundred years before this writing, so in 1907, to locate the logical source of the difficulty with **indirect existence proofs**, as in the above example about natural numbers. The difficulty comes from the law of excluded middle. Moreover, that law leads easily to the related law of double negation: Assume $A \vee \neg A$. If $\neg\neg A$, then A follows by rule *mtp* that is derivable in intuitionistic logic.

The law of excluded middle of classical logic can be interpreted in terms of truth: For any sentence A, either A is true or $\neg A$ is true. If instead of truth we read it in terms of what we can prove, we obtain: For any sentence A, either we can prove A or we can prove $\neg A$. If this were the case, we could in principle decide any claim, but what reason is there for so believing?

There is a practical reason for the use of intuitionistic logic that has nothing to do with possible philosophical motivations. Proofs based on intuitionistic logic are **constructive**: A proof in general shows that a conclusion C follows from some given assumptions $A_1, \dots, A_n$. A constructive proof has an additional property: If the assumptions have been verified, which is by its nature a finitary task, the conclusion can be similarly verified. For example, let the assumptions be mathematical claims with a numerical content. In a specific situation, the correctness of the numerical content can be verified by a finite calculation. Whatever consequences the assumptions have, they can be similarly verified by a finite calculation. In a suggestive language, proofs that are not constructive can lead to claims that are infinitely difficult to resolve, even if the assumptions were resolved. In sum, we have:

Constructive proof maintains the property of computability.

We shall see in the next chapter that this vital property of proofs does not hold in systems in which laws such as those of excluded third or double negation are derivable.

Intuitionistic or constructive logic is today a standard instrument in computer science. Its use in program construction guarantees that the running of programs terminates instead of possibly going on indefinitely.

3.8 Meaning explanations

The philosopher Arthur Prior, known for his contributions to **temporal** logic, fantasized the following connective he called 'tonk', denoted $\mathcal{T}$ in what follows. The rules are:

Table 3.13 The rules of tonkinoise logic

$\dfrac{A}{A\mathcal{T}B}\,\mathcal{T}I$	$\dfrac{B}{A\mathcal{T}B}\,\mathcal{T}I$	$\dfrac{A\mathcal{T}B}{A}\,\mathcal{T}E$	$\dfrac{A\mathcal{T}B}{B}\,\mathcal{T}E$

Here is an example derivation by these rules:

$$\dfrac{\dfrac{A}{A\mathcal{T}B}\,\mathcal{T}I}{B}\,\mathcal{T}E$$

Anything can be inferred from an assumption by these rules. The conclusion is that the rules of logic cannot be chosen in any way one pleases.

We determined in Section 3.4 general forms of elimination rules from the corresponding introduction rules. Gentzen, and Prawitz following him, had come to the idea that the introduction rules give the meanings of the logical operations in terms of proof. More precisely, they give sufficient grounds for asserting a sentence of a given form. Their idea was that the elimination rules should give back these sufficient grounds, so be **inverses** of the introduction rules. The correctness of the rules for a connective can then be controlled by checking if an introduction followed by an elimination really gives back the grounds for the introduction. All of this works well for conjunction. In the more general inversion idea in the beginning of Section 3.4, we have changed the direct grounds for introducing a formula into arbitrary consequences of such grounds. That change brings greater uniformity to the treatment, because the elimination rules are actually determined, not just justified as with Gentzen and Prawitz, from the introduction rules by the general inversion principle. Thus, for example, Prior's introduction rules lead to the elimination rule for disjunction, with no justification for the funny $\mathcal{T}E$-rules.

Implication is a difficult connective in this connection. The introduction rule requires the **existence** of a derivation of B from A, and the inversion principle should correspondingly be formulated in terms of arbitrary consequences of the existence of a derivation. This is possible, but not within propositional logic. Nevertheless, the E-rule we have given for implication lets us define normal derivability as well as the normalization process in a unified manner for all connectives.

The explanation of the meaning of the logical connectives proceeds as follows: First it is laid down what counts as a direct argument for a formula in our system of rules, namely that it is derived by an introduction rule. Let us say that if there are no open assumptions, we have a **categorical** assertion at hand. Its derivation can be converted into a normal derivation. There are no open assumptions and therefore, by the transparent structure that results from the general elimination rules, the last rule has to be an introduction. The formula is consequently derived from its direct grounds. Next this explanation is extended to formulas C derived under assumptions. Let us say that we have a **hypothetical** assertion at hand if C is asserted under the open assumptions $A_1, \ldots, A_n$. The meaning of such an assertion is: Whenever $A_1, \ldots, A_n$ turn into categorical assertions, a categorical assertion of C can be made. All of it hangs on the composability of derivations and on their normalizability in a bounded number of steps of conversion. Without these, there would be no non-circular explanation of the logical connectives in terms of proof.

Notes and exercises to Chapter 3

Natural deduction was discovered by Gerhard Gentzen in 1932 and formed part of his thesis *Untersuchungen über das logische Schliessen* (Investigations into logical inference) that he finished in May 1933 under the supervision of Paul Bernays. The thesis was published in two parts in 1934 and 1935, and an English translation can be found in Gentzen's *Collected Papers.* The proof of normalization is from the same time, but it was not published and remained unknown to logicians who in the meanwhile produced new proofs of the result. An English translation of Gentzen's proof is published in his 'The normalization of derivations' (2008).

Section 3.5(b) presented the logical fallacy by which the derivability of $A \supset B$ is claimed to follow, if the derivability of B follows from the derivability of A. A discussion, with examples of this error in various guises drawn from articles and books in logic, is found in Hakli and Negri (2012).

1. Translate the linear derivation of $(A \supset (A \supset B)) \supset (A \supset B)$ in Section 2.4 into tree form.

2. Translate some of the linear derivations of exercise 3 in Chapter 2 into tree form.

3. Give tree derivations of the following:
   a. $A \vee A \vdash A$, $A \vdash A \vee A$
   b. $A \vee B \vdash \neg(\neg A \& \neg B)$
   c. $\neg A \vee B \vdash A \supset B$
   d. $\vdash \neg(A \& \neg A)$
   e. $\vdash \neg\neg(A \vee \neg A)$

4. Conjunctive and disjunctive conditions:
   a. Why is $(A \supset C) \& (B \supset C)$ not derivable from $A \& B \supset C$?
   b. Show $(A \vee B \supset C) \supset (A \& B \supset C)$.

5. Work out the second detour conversion for disjunction.

6. Give a derivation of the formula $A \vee \neg B \supset (B \supset A)$. By example 3.3(c), $A \vee B \supset C \vdash A \supset C$ and $A \vee B \supset C \vdash B \supset C$. Apply these derivability relations to the formula.

7. Complete the translation from negation as a primitive into negation as a defined notion of Section 3.5(c). Translate the derivations at the end of the section, with repeated instances of $\neg I$, $\neg E$, so that they use the defined notion of negation. Make the detour conversions that appear in translation. Check that the simplified derivations that end the section translate into the converted derivations.

8. Try to show that the following implications are not derivable in intuitionistic logic:
   a. $\neg(\neg A \vee \neg B) \supset A \& B$
   b. $\neg(\neg A \& \neg B) \supset A \vee B$
   c. $\neg(A \& \neg B) \supset (A \supset B)$

9. Define a translation from tree derivations to linear derivations.

4 | Proof search

The construction of derivations in Gentzen's tree form is awkward: One would have to know, more or less, how the final tree has to look before one can start. So one looks at the assumptions, and one looks at the conclusion, and one tries to figure a way from the former to the latter. The good aspects of the tree form include that once you have the tree, its structure is transparent.

We shall modify a bit natural deduction, to get a logical calculus that supports proof search better. The starting point is the change in Section 3.4 in which all E-rules were written with an arbitrary conclusion. Consider any formula C in a derivation tree. The assumptions it depends on can be listed, and let them be $A_1, \dots, A_n$. If you take the part of the tree that is determined by the chosen formula C, you get a **subderivation** of the original derivation. To be precise, to get a correct derivation tree you have to delete the labels from above those assumptions that have not been closed at the stage in which formula C is concluded. The subderivation establishes a derivability relation, namely $A_1, \dots, A_n \vdash C$.

It will be convenient to have a name for the thing a derivability relation establishes: Expressions of the form $A_1, \dots, A_n \vdash C$ are called **sequents.** The name stems from the arrangement of formulas in a sequence in the **antecedent**, left part of a sequent. The right part of a sequent is its **succedent**. The codification of logical rules in this notation is called **sequent calculus**, the second of Gentzen's great discoveries in pure logic. We shall see that the notation lends perfect support for the construction of derivations: Given that we have to establish a derivability relation $A_1, \dots, A_n \vdash C$, we can apply the logical rules to write a premiss or premisses from which $A_1, \dots, A_n \vdash C$ can be concluded, then to repeat this procedure, until this **root-first proof search** terminates. A powerful method results once we have established termination, after which derivations as well as proofs of underivability through failed proof search are easy to find.

4.1 Naturally growing trees

In general, if our task is to establish a derivability relation $\Gamma \vdash C$, with Γ a list of assumptions, we can keep proof search **local** by a suitable notation for derivations. If the conclusion under the assumptions Γ is a conjunction

$A \& B$, it is sufficient to establish $\Gamma \vdash A$ and $\Gamma \vdash B$. If it is an implication $A \supset B$, it is sufficient to establish $A, \Gamma \vdash B$. If it is a disjunction $A \vee B$, it is sufficient to establish one of $\Gamma \vdash A$ or $\Gamma \vdash B$.

Instead of analysing the conclusion, we can work on the assumption side. If one assumption is a conjunction $A \& B$ and the derivability relation to be established $A \& B, \Gamma \vdash C$, it is sufficient to establish $A, B, \Gamma \vdash C$. If an assumption is a disjunction and the derivability relation $A \vee B, \Gamma \vdash C$, it is sufficient to establish $A, \Gamma \vdash C$ and $B, \Gamma \vdash C$. With an implication, to establish $A \supset B, \Gamma \vdash C$, it is sufficient to establish $A \supset B, \Gamma \vdash A$ and $B, \Gamma \vdash C$. Finally, if we have to establish $\bot, \Gamma \vdash C$, it is already done, and the same if we have to establish $A, \Gamma \vdash A$.

Let us encode the above as a system of rules. We write the names of the rules so that it is seen if they are rules of natural deduction or of sequent calculus for root-first proof search. Instead of writing I and E for introduction and elimination, we write R and L. The former give as a conclusion a sequent with the **principal** formula, i.e., the formula with the connective of the rule, at right of the derivability relation, the latter with the principal formula at left. The components of the principal formula in the premisses are the **active** formulas of a rule.

Table 4.1 Rules for root-first proof search

$$\frac{\Gamma \vdash A \quad \Gamma \vdash B}{\Gamma \vdash A \& B}{\scriptstyle R\&} \qquad \frac{\Gamma \vdash A}{\Gamma \vdash A \vee B}{\scriptstyle R\vee_1} \qquad \frac{\Gamma \vdash B}{\Gamma \vdash A \vee B}{\scriptstyle R\vee_2} \qquad \frac{A, \Gamma \vdash B}{\Gamma \vdash A \supset B}{\scriptstyle R\supset}$$

$$\frac{A, B, \Gamma \vdash C}{A \& B, \Gamma \vdash C}{\scriptstyle L\&} \qquad \frac{A, \Gamma \vdash C \quad B, \Gamma \vdash C}{A \vee B, \Gamma \vdash C}{\scriptstyle L\vee} \qquad \frac{A \supset B, \Gamma \vdash A \quad B, \Gamma \vdash C}{A \supset B, \Gamma \vdash C}{\scriptstyle L\supset}$$

These rules are closely related to the rules of natural deduction of Section 3.4. We say that we construct derivations with them, and that a sequent $\Gamma \vdash C$ is derivable if it is the last line of a derivation constructed by the rules. This derivability is of a higher level as compared to derivability in the sense of the derivability relation for which the symbol $\vdash$ is used. The terminology of sequents avoids the double meaning of derivability. The last sequent in a derivation is the **endsequent** of the derivation.

The R-rules for conjunction and implication are **invertible**: Whenever a sequent of the form $\Gamma \vdash A \& B$ is derivable, the sequents $\Gamma \vdash A$ and $\Gamma \vdash B$ are derivable. Note that the last rule in the derivation of $\Gamma \vdash A \& B$ **need not be** $R\&$. By invertibility, however, $\Gamma \vdash A \& B$ **can be** concluded by $R\&$. Similarly, whenever a sequent $\Gamma \vdash A \supset B$ is derivable, the sequent $A, \Gamma \vdash B$ is derivable. Rules $R\vee_1$ and $R\vee_2$ are not invertible. On the antecedent side, rules $L\&$ and $L\vee$ are invertible. With rule $L\supset$, if a sequent of the form $A \supset B, \Gamma \vdash C$ is derivable, the sequent $B, \Gamma \vdash C$ that matches the second premiss of the rule is derivable, but the sequent $A \supset B, \Gamma \vdash A$ need not be derivable.

It may seem strange that the principal formula of rule $L\supset$ is repeated in the first premiss. The repetition can be justified, though not motivated, as follows: If in an endsequent it is allowed to assume $A \supset B$, the same assumption must be allowed anywhere.

To show that a sequent is derivable, we start **decomposing** formulas in it in the root-first direction. It is best to decompose first formulas that correspond to invertible rules, first those that have one premiss. Successive decomposition amounts to the root-first construction of branches in a derivation tree. A branch ends when we reach an **initial sequent**, either one that has the same formula as an assumption and conclusion, or one that has $\bot$ as an assumption.

Table 4.2 Initial sequents

$A, \Gamma \vdash A$	$\bot, \Gamma \vdash C$

A sequent of the form $A, \Gamma \vdash A$ corresponds to the making of an assumption in derivations of Gentzen's tree form. A sequent of the form $\bot, \Gamma \vdash C$ corresponds to rule $\bot E$. That is why it is sometimes written as a limiting case of a rule when the number of premisses is 0.

Some examples will show how the calculus of sequents works. It must be kept in mind that the derivations are best read the way they are constructed, i.e., starting from the root.

a. $(A \supset B) \& (B \supset C) \vdash A \supset C$

$$\dfrac{\dfrac{\dfrac{A \supset B, B \supset C, A \vdash A \qquad \dfrac{B, B \supset C, A \vdash B \quad C, B, A \vdash C}{B, B \supset C, A \vdash C}\,{\scriptstyle L\supset}}{A \supset B, B \supset C, A \vdash C}\,{\scriptstyle L\supset}}{(A \supset B) \& (B \supset C), A \vdash C}\,{\scriptstyle L\&}}{(A \supset B) \& (B \supset C) \vdash A \supset C}\,{\scriptstyle R\supset}$$

b. $\neg\neg(A \vee \neg A)$

$$\dfrac{\dfrac{\dfrac{\dfrac{\dfrac{\dfrac{\neg(A \vee \neg A), A \vdash A}{\neg(A \vee \neg A), A \vdash A \vee \neg A}\,{\scriptstyle R\vee_1} \qquad \bot, A \vdash \bot}{\neg(A \vee \neg A), A \vdash \bot}\,{\scriptstyle L\supset}}{\neg(A \vee \neg A) \vdash \neg A}\,{\scriptstyle R\supset}}{\neg(A \vee \neg A) \vdash A \vee \neg A}\,{\scriptstyle R\vee_2} \qquad \bot \vdash \bot}{\neg(A \vee \neg A) \vdash \bot}\,{\scriptstyle L\supset}}{\vdash \neg\neg(A \vee \neg A)}\,{\scriptstyle R\supset}$$

This second example goes through smoothly, but the same cannot be said of the corresponding derivation in Gentzen's natural deduction, as in exercise 3(e) of the previous chapter.

The first, i.e. downmost, rule in example (b) has to be $R\supset$, the second $L\supset$. At this point, we notice that if the principal formula $\neg(A \vee \neg A)$ had not been repeated in the left premiss, that premiss would be $\vdash A \vee \neg A$ which is underivable in intuitionistic logic. There were no choices up to that point, so that without repetition the calculus would not derive what it should derive. The first choice in the proof search was when we came to the said left premiss. Had we tried again rule $L\supset$, the part of derivation would have become:

$$\frac{\neg(A \vee \neg A) \vdash A \vee \neg A \qquad \bot \vdash A \vee \neg A}{\neg(A \vee \neg A) \vdash A \vee \neg A}{\scriptstyle L\supset}$$

The left premiss is identical to the conclusion, so nothing was gained by the step, but only a **loop** produced. Therefore:

No derivation branch must contain the same sequent twice.

The rules of the calculus of sequents are **local**, and therefore loops can be eliminated by deleting the part of derivation between the two occurrences. In this case, the second premiss $\bot \vdash A \vee \neg A$ and its (degenerate) derivation would have fallen off the derivation tree.

With loops forbidden, the only choice for the third rule root first is $R\vee_1$ or $R\vee_2$. Then, if we try $R\vee_1$, the premiss is $\neg(A \vee \neg A) \vdash A$ and the only applicable rule, namely $L\supset$, will again give a loop:

$$\frac{\dfrac{\neg(A \vee \neg A) \vdash A \vee \neg A \qquad \bot \vdash A}{\neg(A \vee \neg A) \vdash A}{\scriptstyle L\supset}}{\neg(A \vee \neg A) \vdash A \vee \neg A}{\scriptstyle R\vee_1}$$

The identical sequents are separated by two steps. The loop is deleted by continuing from the left premiss of $L\supset$ in the way the original derivation continues from the conclusion of $R\vee_1$. A whole derivation branch falls off, as above.

With loops forbidden, the only rule applicable above the two last ones in the derivation is $R\vee_2$. If to the premiss $\neg(A \vee \neg A) \vdash \neg A$ rule $L\supset$ is applied, a loop is again produced. Therefore the rule to be applied is $R\supset$ and the premiss is $\neg(A \vee \neg A), A \vdash A \vee \neg A$. At this point, it is seen that rule $R\vee_1$ gives an initial sequent as a premiss.

If instead of $R\vee_1$ as an uppermost rule we try $R\vee_2$, we get the sequent $\neg(A \vee \neg A), A \vdash \neg A$ to which $R\supset$ has to be applied lest a loop be produced. The result is $\neg(A \vee \neg A), A, A \vdash \bot$. Now rule $L\supset$ gives as the first premiss $\neg(A \vee \neg A), A, A \vdash A \vee \neg A$, and it is the only applicable rule. We get a sequent that is just like an earlier one, except for the **duplication** of the formula A in the antecedent part. Obviously nothing is gained by the duplication, but we note that proof search can fail for the reason that the search can go on producing multiplications forever. To avoid the phenomenon, we show that no rule instances are needed in which a sequent is produced that is exactly like some previous sequent in the same branch, save for possible duplications.

4.2 Invertibility

The application of invertible rules will not lead proof search to a dead end, if the sequent to be derived really is derivable. We prove next the invertibilities mentioned in the previous section. The proof uses a principle of structural induction on derivations, explained in Section 13.1.

Inversion lemma.

(i) *If* $\Gamma \vdash A \& B$, *then* $\Gamma \vdash A$ *and* $\Gamma \vdash B$.
(ii) *If* $A \& B, \Gamma \vdash C$, *then* $A, B, \Gamma \vdash C$.
(iii) *If* $A \vee B, \Gamma \vdash C$, *then* $A, \Gamma \vdash C$ *and* $B, \Gamma \vdash C$.
(iv) *If* $\Gamma \vdash A \supset B$, *then* $A, \Gamma \vdash B$.
(v) *If* $A \supset B, \Gamma \vdash C$, *then* $B, \Gamma \vdash C$.

Proof. The proof is by induction on the height (greatest number of successive rules) in a derivation.

(i) Inversion of $\Gamma \vdash A \& B$.

Base case: If $\Gamma \vdash A \& B$ is an initial sequent, either $A \& B$ is in Γ or $\bot$ is in Γ. In the latter case $\Gamma \vdash A$ and $\Gamma \vdash B$ are initial sequents and therefore derivable. In the former case, $\Gamma \equiv A \& B, \Gamma'$ and $\Gamma \vdash A$ and $\Gamma \vdash B$ are derivable:

$$\frac{A, B, \Gamma' \vdash A}{A \& B, \Gamma' \vdash A}\,L\& \qquad \frac{A, B, \Gamma' \vdash B}{A \& B, \Gamma' \vdash B}\,L\&$$

Inductive case: If $A \& B$ is principal in the last rule of the derivation of $\Gamma \vdash A \& B$, the premisses give the inversions. Otherwise it is not principal.

Apply the inductive hypothesis to the premisses of the last rule, then the last rule.

(ii) Inversion of $A \& B, \Gamma \vdash C$.

Base case: If $A \& B, \Gamma \vdash C$ is an initial sequent, either C is in Γ or $\bot$ is in Γ and also $A, B, \Gamma \vdash C$ is an initial sequent. Else $C \equiv A \& B$ and $A, B, \Gamma \vdash A \& B$ is derived by:

$$\frac{A, B, \Gamma \vdash A \quad A, B, \Gamma \vdash B}{A, B, \Gamma \vdash A \& B}{\scriptstyle R\&}$$

Inductive case: As in (i).

(iii) Inversion of $A \vee B, \Gamma \vdash C$. This goes through in a way similar to (ii).

(iv) Inversion of $\Gamma \vdash A \supset B$. This goes through in a way similar to (i).

(v) Inversion of $A \supset B, \Gamma \vdash C$.

Base case: As in (ii) except when $C \equiv A \supset B$. Then $B, \Gamma \vdash A \supset B$ is derived by:

$$\frac{A, B, \Gamma \vdash B}{B, \Gamma \vdash A \supset B}{\scriptstyle R\supset}$$

Inductive case: As in (ii). QED.

Contrary to invertible rules, the root-first application of non-invertible rules can lead proof search to a dead end. In the simplest case, $A \vdash A \vee B$ is derivable, but the application of rule $R\vee_2$ gives $A \vdash B$ that is not derivable, and similarly for $B \vdash A \vee B$ and rule $R\vee_1$. For the case of rule $L\supset$, the sequent $A \supset B \vdash A \supset B$ gives an example of failure of invertibility: It is an initial sequent and therefore derivable. Application of rule $L\supset$ gives:

$$\frac{A \supset B \vdash A \quad B \vdash A \supset B}{A \supset B \vdash A \supset B}{\scriptstyle L\supset}$$

The only rule that can be applied to the left premiss is again $L\supset$ but it gives a loop. Therefore $A \supset B \vdash A$ is not derivable and rule $L\supset$ not invertible with respect to its first premiss.

Looking at the invertible rules, we notice that each premiss has a number of connectives that is at least one less the number of connectives in the conclusion. Invertible rules thus simplify the task of finding a derivation. We shall see in the next chapter how classical propositional logic improves the situation even further: Each rule is invertible, each premiss is simpler than the conclusion, and proof search terminates irrespective of the order of application of the rules.

4.3 Translation to sequent calculus

We shall now define a translation from natural deduction to sequent calculus. We start from the rules of Section 3.4, those that had an arbitrary conclusion in the E-rules. Looking at the rules of sequent calculus of Table 4.1, we notice that the principal formulas in L-rules, so on the left side, are always among the assumptions. The corresponding property for natural derivations is that the major premisses of E-rules are assumptions, which means that a derivation is normal. Consequently, the translation we define relates the logical rules of sequent calculus to normal instances of the natural rules.

In those rules of natural deduction that have two or three premisses, derivations of these premisses are combined together to yield the conclusion. We have, say, derivations of the premisses $\Gamma \vdash A$ and $\Delta \vdash B$ and rule $R\&$ gives the conclusion $\Gamma, \Delta \vdash A \& B$. The collections of open assumptions or **contexts** of A and B, respectively, are **independent** and put together to form the collection of assumptions Γ, Δ of the conclusion. In sequent calculus, rules that are applied root first have instead **shared** contexts: All open assumptions of a conclusion $A \& B$ of rule $R\&$ are repeated in both premisses, and similarly for the other logical rules.

The possibility of vacuous and multiple discharge in natural deduction calls for some adjustments in the translation. If rule $R\supset$ is applied, the derivability relation that is established is of the form $\Gamma \vdash A \supset B$ with Γ the collection of open assumptions. The premiss is, with a multiple, simple, and vacuous discharge, respectively:

$$A, \dots, A, \Gamma \vdash B \qquad A, \Gamma \vdash B \qquad \Gamma \vdash B$$

There are two ways of handling the situation with vacuous and multiple discharge. In the translation, the multiple occurrence of A in the antecedent of the former can be **contracted** into one occurrence. With a vacuous discharge, the missing occurrence of A in the antecedent can be fixed in the translation by **weakening** the antecedent Γ into A, Γ. Multiple and vacuous closing of assumptions in the E-rules presents a similar situation and is exemplified by:

$$\dfrac{A^m, B^n, \Gamma \vdash C}{A \& B, \Gamma \vdash C}{\scriptstyle L\&}$$

A formally correct way to deal with weakening and contraction is to add explicit **structural rules** to sequent calculus:

$$\dfrac{\Gamma \vdash C}{A, \Gamma \vdash C}{\scriptstyle Wk} \qquad \dfrac{A, A, \Gamma \vdash C}{A, \Gamma \vdash C}{\scriptstyle Ctr}$$

A second, alternative way to deal with the translation of vacuous and multiple discharge is to allow an arbitrary number of active formulas in the antecedents of sequent rules.

With these preparations, we can now give the translation manual from normal derivations in natural deduction to ones in sequent calculus:

Table 4.3 Translation procedure into sequent calculus

1. *Collect the open assumptions $A_1, \dots A_n$ of the endformula C into a sequent $A_1, \dots, A_n \vdash C$ and draw the inference line with the rule symbol above it.*
2. *If the last rule was an I-rule, repeat 1 above the inference line for the premiss or premisses. With more than one premiss, make the contexts equal by suitably adding formulas.*
3. *If the last rule was an E-rule, repeat 1 for the minor premisses. With $L\supset$, add its major premiss in the antecedent of the sequent that forms the left premiss.*
4. *With rules that close assumptions, contract multiple occurrences of the assumption formula into one occurrence. With missing assumption formulas, weaken by adding one occurrence.*

Let us take as an example the first one from Section 3.1, namely:

$$(A \supset B) \& (B \supset C) \vdash A \supset C$$

Its derivation in Section 3.1 is normal in the sense of Gentzen's original calculus, but it is not normal in the final sense of the word, as in Section 3.4, because there are two instances of rule $L\supset$ in which the major premiss has been derived. We shall therefore first give a normal derivation with the general E-rules:

$$\dfrac{\dfrac{(A \supset B) \& (B \supset C) \qquad \dfrac{\overset{3}{B \supset C} \qquad \dfrac{\overset{3}{A \supset B} \quad \overset{4}{A} \quad \overset{1}{B}}{B}{\scriptstyle \supset E,1} \qquad \overset{2}{C}}{C}{\scriptstyle \supset E,2}}{C}{\scriptstyle \& E,3}}{A \supset C}{\scriptstyle \supset I,4}$$

The root-first translation gives as a last step $R\supset$:

$$\frac{}{(A \supset B)\,\&\,(B \supset C) \vdash A \supset C}\,R\supset$$

On the line above, C is concluded by rule $\&E$ and assumption A has become open, so it has to be added to the list in the antecedent:

$$\frac{}{(A \supset B)\,\&\,(B \supset C),\, A \vdash C}\,L\&$$

Next we have the derivation of the minor premiss C from the assumptions $A \supset B$, $B \supset C$, and A:

$$\frac{}{A \supset B,\, B \supset C,\, A \vdash C}\,L\supset$$

This is an L-rule with the two minor premisses B and C in the natural derivation and we get two sequents, with the major premiss $B \supset C$ added in the antecedent of the first:

$$\frac{}{B \supset C,\, A \supset B,\, A \vdash B}\,L\supset \quad \text{and} \quad C,\, A \supset B,\, A \vdash C$$

The final step gives the premisses of $L\supset$:

$$B \supset C,\, A \supset B,\, A \vdash A \quad \text{and} \quad B,\, B \supset C,\, A \vdash B$$

Putting it all together, we have:

$$\cfrac{\cfrac{\cfrac{\cfrac{B \supset C,\, A \supset B,\, A \vdash A \qquad B,\, B \supset C,\, A \vdash B}{B \supset C,\, A \supset B,\, A \vdash B}\,L\supset \qquad C,\, A \supset B,\, A \vdash C}{A \supset B,\, B \supset C,\, A \vdash C}\,L\supset}{(A \supset B)\,\&\,(B \supset C),\, A \vdash C}\,L\&}{(A \supset B)\,\&\,(B \supset C) \vdash A \supset C}\,R\supset$$

All discharges were simple.

Let us take examples in which clause (4) of the translation manual is put into use:

$$\cfrac{\cfrac{\cfrac{\overset{4}{A \supset (A \supset B)} \quad \overset{3}{A} \quad \cfrac{\overset{2}{A \supset B} \quad \overset{3}{A} \quad \overset{1}{B}}{B}\,{\supset}E,1}{B}\,{\supset}E,2}{A \supset B}\,{\supset}I,3}{(A \supset (A \supset B)) \supset (A \supset B)}\,{\supset}I,4$$

The translation gives first:

$$\cfrac{\cfrac{A,\, A,\, A \supset (A \supset B) \vdash B}{A \supset (A \supset B) \vdash A \supset B}\,R\supset}{\vdash (A \supset (A \supset B)) \supset (A \supset B)}\,R\supset$$

The duplication of A in the premiss must be first contracted by clause (4), after which the translation continues:

$$\dfrac{A, A \supset (A \supset B) \vdash A \qquad \dfrac{A \supset B, A \vdash A \qquad B, A \vdash B}{A \supset B, A \vdash B}\,L\supset}{A, A \supset (A \supset B) \vdash B}\,L\supset$$

Weakening of the antecedent is put into use in the translation of:

$$\dfrac{\dfrac{\overset{1}{A}}{B \supset A}\,\supset I}{(A \supset (B \supset A)}\,\supset I,1$$

The translation gives:

$$\dfrac{\dfrac{A, B \vdash A}{A \vdash B \supset A}\,R\supset}{\vdash A \supset (B \supset A)}\,R\supset$$

One occurrence of B has been added in the topsequent, because B was vacuously discharged in the natural derivation.

The translation defined by Table 4.3 is straightforward. There is no difficulty in defining a translation in the other direction, from sequent calculus to natural deduction:

Table 4.4 Translation from sequent calculus to natural deduction

1. *Write the right side of the endsequent as the endformula and above it the inference line and rule.*
2. *For rules other than $R\supset$ and the L-rules, continue from the premisses.*
3. *For $R\supset$ with conclusion $A \supset B$, add a label above A in the sequent that forms the premiss. Add similarly labels for the L-rules and write out their major premisses, then continue from the minor premisses. Labelled formulas keep their labels in each occurrence.*
4. *When you come to an initial sequent $\bot, \Gamma \vdash C$, write an instance of rule $\bot E$. In the other case of an initial sequent, the form will be $A, \Gamma \vdash A$ or $\overset{n}{A}, \Gamma \vdash A$. Write A and $\overset{n}{A}$, respectively.*

We conclude that there cannot be any fundamental difference between normal derivations in natural deduction with the modified E-rules of Section 3.4 and derivations in the sequent calculus. In fact, as seen from the translations, we have:

Isomorphism of natural and sequent derivations. *The order of application of the logical rules is preserved in the translations between normal natural* *derivations with general E-rules and sequent derivations with logical rules and weakening and contraction.*

The interest in Gentzen's tree form for natural deduction lies in the ease with which non-normalities can be eliminated by detour and permutation conversions, as seen in Sections 3.2 and 3.3. It is possible to generalize the sequent calculus of this chapter so that what corresponds to non-normalities is seen in sequent calculus derivations, as in Section 13.4. It then happens that formulas in premisses are not necessarily found in the conclusion. If we instead use only normal instances, as in the rules of Table 4.1, we see at once that in the translation to sequent calculus, each formula in the premisses of a rule is also a formula in the conclusion. Therefore the subformula property is immediate:

Subformula property for sequent calculus. *Each formula in a derivation with the logical rules of sequent calculus is a subformula in the endsequent.*

4.4 Unprovability through failed proof search

Sequent calculus offers a good possibility for **exhaustive proof search** in propositional logic: We can check through all the possibilities for making a derivation. If none of them worked, i.e., if each had at least one branch in which no rule applied and no initial sequent was reached, the given sequent is underivable. The symbol $\nvdash$ is used for underivability.

Examples of exhaustive proof search:

a. $\nvdash A$

No rule can be applied.

b. $\nvdash \bot$

No rule can be applied, so that sequent calculus is consistent.

c. $\nvdash A \vee \neg A$

Rule $R\vee_1$ gives example (a), rule $R\vee_2$ gives $\vdash \neg A$, so by $R\supset$, the sequent to be derived is $A \vdash \bot$, and now no rule applies.

d. $\nvdash \neg\neg A \supset A$

Rule $R\supset$ gives $\neg\neg A \vdash A$. Next rule $L\supset$ gives $\neg\neg A \vdash \neg A$ and $\bot \vdash \bot$. There are two possibilities for $\neg\neg A \vdash \neg A$ and we can without limitation apply rule $R\supset$ because it is invertible, to get $\neg\neg A,\ A \vdash \bot$. Rule $L\supset$ gives the premisses $\neg\neg A,\ A \vdash \neg A$ and $\bot,\ A \vdash \bot$. Now $R\supset$ gives $\neg\neg A,\ A,\ A \vdash \bot$, with a duplication of A in an antecedent. Further repetition of the procedure just keeps multiplying the number of occurrences of A in the antecedent, with no end.

e. $\nvdash \neg(\neg A \& \neg B) \supset A \vee B$

An exhaustive proof search would involve a lot of writing. We show how a leftmost branch in what is the only reasonable line of attack leads to a multiplication of the same formula in an antecedent:

$$
\dfrac{\dfrac{\dfrac{\dfrac{\dfrac{\dfrac{\dfrac{\vdots}{\neg(\neg A \& \neg B),\ A,\ A \vdash \bot}}{\neg(\neg A \& \neg B),\ A \vdash \neg A}{\scriptstyle R\supset} \quad \dots}{\neg(\neg A \& \neg B),\ A \vdash \neg A \& \neg B}{\scriptstyle R\&} \quad \dots}{\neg(\neg A \& \neg B),\ A \vdash \bot}{\scriptstyle L\supset}}{\neg(\neg A \& \neg B) \vdash \neg A}{\scriptstyle R\supset} \quad \dots}{\neg(\neg A \& \neg B) \vdash \neg A \& \neg B}{\scriptstyle R\&} \qquad \bot \vdash A \vee B}{\neg(\neg A \& \neg B) \vdash A \vee B}{\scriptstyle L\supset}}{\vdash \neg(\neg A \& \neg B) \supset A \vee B}{\scriptstyle R\supset}
$$

With $R\&$, the third-to-last step, the rule is invertible and both branches have to lead to initial sequents, so we took the left one. The only real choice in the proof search was with the second-to-last rule that could have been $\vee R$. With the next step, all traces of this choice would have disappeared, as in:

$$
\dfrac{\dfrac{\dfrac{\neg(\neg A \& \neg B) \vdash \neg A \& \neg B \quad \bot \vdash A}{\neg(\neg A \& \neg B) \vdash A}{\scriptstyle L\supset}}{\neg(\neg A \& \neg B) \vdash A \vee B}{\scriptstyle R\vee_1}}{\vdash \neg(\neg A \& \neg B) \supset A \vee B}{\scriptstyle R\supset}
$$

The left premiss is found also in the first attempt.

The underivability of connections between the connectives &, $\vee$, and $\supset$, mentioned in Section 3.7(b), and not straightforward to argue for in natural deduction by exercise 8 of Chapter 3, can now be carried through without great difficulties and is left as an exercise.

Let a sequent $\Gamma \vdash A$ contain 50 connectives in an arrangement in which each applicable rule has two premisses. Up to 50 applications of rules are possible along each branch of a derivation, so that the number of leaves of the derivation tree is at most 2^{50}. There is not enough paper or computer memory to write down such derivations. Let us assume further that $\Gamma \nvdash A$.

Every attempt at a derivation will lead to at least one branch that ends in an unanalysable sequent that is not an initial sequent or else the branch goes on forever. Given an underivable sequent, is there a fast method for showing that at least one branch leads to a failure of proof search? Similarly, given a derivable sequent, is there a fast method for showing that some derivation tree has only initial sequents as leaves? There are positive answers to limited classes of formulas, but the general answer to this '$P = NP$-question' is unknown. Most researchers tend to think that there is no general method for speeding up proof search in propositional logic from the exponential growth that is the worst case if whole proof trees have to be constructed.

4.5 Termination of proof search

Apart from the practical impossibility of constructing derivations with a great number of atomic formulas and connectives, there is the question of principle of termination of proof search. Looking at the rules in Table 4.1, we notice that all formulas in premisses are subformulas of formulas that are also found in the conclusion. Thus, no new formulas can surface in a root-first proof search and the number of distinct formulas that can occur in proof search is bounded: Possible non-termination can occur only as a result of multiplication of formulas in the antecedents of premisses.

The premisses are simpler than the conclusion in all the rules except possibly in the left premiss of rule $L\supset$. That is the only source of non-termination. Rules other than $L\supset$ can produce duplication, if an active formula had another occurrence in the antecedent. This source of duplication comes to an end.

There are at least two solutions to the problem of non-termination. One is to show that if a sequent with a duplication is derivable, as in $A, A, \Gamma \vdash C$, then also $A, \Gamma \vdash C$ is derivable:

Theorem. Admissibility of contraction. *Whenever a sequent of the form $A, A, \Gamma \vdash C$ is derivable, also $A, \Gamma \vdash C$ is derivable.*

Proof. The proof is by induction on the length of the contraction formula, with a subinduction on the height of derivation of the derivable sequent with a duplication. It is based on the inversion lemma of Section 4.2.

Base case (contraction with a simple formula): The derivable sequent is $P, P, \Gamma \vdash C$ with P a simple formula.

Subinduction on height of derivation: If $P,\ P, \Gamma \vdash C$ is an initial sequent, also $P, \Gamma \vdash C$ is. Since P is not principal in any rule, the inductive case on height of derivation for a simple formula goes through as in the inversion lemma.

Inductive case (contraction with a compound formula): There are three forms of compound formulas:

(i) Conjunction: Let the sequent be $A \& B,\ A \& B, \Gamma \vdash C$. Base case of subinduction on height of derivation as above. Inductive case: If $A \& B$ is not principal, apply the inductive hypothesis to the premisses of the last rule for the derivability of $A \& B, \Gamma \vdash C$. If instead $A \& B$ is principal, the last rule is $L\&$ and the premiss $A,\ B,\ A \& B, \Gamma \vdash C$. By the inversion lemma, $A,\ B,\ A,\ B, \Gamma \vdash C$ is derivable. Now apply contraction to the shorter formulas A and B to conclude the derivability of $A,\ B, \Gamma \vdash C$. Finally, rule $L\&$ gives $A \& B, \Gamma \vdash C$.

(ii) Disjunction: Let the sequent be $A \vee B,\ A \vee B, \Gamma \vdash C$. This goes through analogously to (i).

(iii) Implication: Let the sequent be $A \supset B,\ A \supset B, \Gamma \vdash C$. Things go through as before except when $A \supset B$ is principal. Then the last step is:

$$\frac{A \supset B,\ A \supset B, \Gamma \vdash A \quad B,\ A \supset B, \Gamma \vdash C}{A \supset B,\ A \supset B, \Gamma \vdash C}\ {\scriptstyle L\supset}$$

Apply the inductive hypothesis to the left premiss for the derivability of $A \supset B, \Gamma \vdash A$. Apply the inversion lemma to the right premiss for the derivability of $B,\ B, \Gamma \vdash C$, then the inductive hypothesis for the derivability of $B, \Gamma \vdash C$. Now rule $L \supset$ gives:

$$\frac{A \supset B, \Gamma \vdash A \quad B, \Gamma \vdash C}{A \supset B, \Gamma \vdash C}\ {\scriptstyle L\supset}$$

By the above, any formula in an antecedent of a sequent can be contracted with derivability of the sequent maintained. QED.

As noted above, root-first proof search in our sequent calculus for propositional logic produces sequents in which all formulas are known from the endsequent. Therefore the number of distinct formulas in a derivation is bounded and possible non-termination can occur only through the duplication of formulas. Let now $A,\ A, \Gamma \vdash C$ be derivable. By the admissibility of contraction, also $A, \Gamma \vdash C$ is derivable. Take a derivation of the latter and you obtain a derivation of $A, A, \Gamma \vdash C$ by adding one copy of A to each antecedent of each sequent in the derivation. We can therefore make the following:

Important observation. *If* $A, A, \Gamma \vdash C$ *is derivable, it has a derivation in which one copy of* A *is never active in the derivation.*

Duplications are useless in proof search because one never needs to use more than one of the copies of a duplicated formula. We can therefore put a 'ban' on rule instances that give as a premiss a sequent that is just like some previous sequent below, except for a duplication. Note that this restriction on proof search does not eliminate all duplications: For example, rule $L\&$ can be applied root-first to the sequent $A \& A \vdash A$.

With a bounded number of distinct formulas, it can happen that each branch in a proof search for a sequent $\Gamma \vdash C$ terminates with an initial sequent. If this is so, $\Gamma \vdash C$ is derivable. Otherwise there are branches that terminate in sequents that have only simple formulas but are not initial sequents, or in sequents that give loops, possibly after some duplications have been contracted. If this is so, $\Gamma \vdash C$ is not derivable. Therefore we have:

Decidability of derivability in intuitionistic propositional logic. *The derivability of a formula* C *from given assumptions* $A_1, \dots, A_n$ *in intuitionistic propositional logic is decidable.*

Notes and exercises to Chapter 4

The single-succedent sequent calculus of proof search of Table 4.1 is a relatively recent invention: Building on the work of Albert Dragalin (1978) on the invertibility of logical rules in sequent calculi, Anne Troelstra worked out the details of the proof theory of this 'contraction-free' calculus in the book *Basic Proof Theory* (2000).

The calculus of this chapter can be so modified that the same sequents are derivable as before, but that proof search terminates in a predictable number of steps without possible looping along branches. The only changes are, as can be expected, with rule $L\supset$. The modified rule is divided into four parts, according to the form of the principal formula, for which see *Structural Proof Theory*, section 5.5.

1. Give root-first derivations of the following:

   a. $(A \supset B) \supset (\neg B \supset \neg A)$

   b. $A \vee B \supset \neg(\neg A \& \neg B)$

c. $\neg(A \supset B) \supset \neg\neg A \& \neg B$
d. $\neg A \vee B \supset (A \supset B)$
e. $\neg\neg(A \& B) \supset \neg\neg A \& \neg\neg B$

2. Show the following in intuitionistic logic:
   a. $\nvdash \neg(\neg A \vee \neg B) \supset A \& B$
   b. $\nvdash \neg(A \& \neg B) \supset (A \supset B)$

3. Show that $(A \vee B) \& \neg B \supset A$ is underivable in minimal logic.

4. Show that the following are not derivable in intuitionistic logic:
   a. $(\neg\neg A \supset A) \supset (A \vee \neg A)$
   b. $(\neg A \vee \neg\neg A) \supset (A \vee \neg A)$

5. **Independence of the intuitionistic connectives.** With P and Q atomic formulas and C &-free, show $\nvdash P \& Q \supset\subset C$. Show $\nvdash (P \supset Q) \supset\subset C$ if C contains no $\supset$ and $\nvdash P \vee Q \supset\subset C$ if C contains no $\vee$. Warning: the last requires the use of Harrop's theorem for which see *Structural Proof Theory*, section 2.5.

5 | Classical natural deduction

5.1 Indirect proof

We showed in Section 3.7(a) that the law of double negation, $\neg\neg A \supset A$, is not derivable in intuitionistic logic. The proof of underivability was done with formal detail in Section 4.4, example (d). The difference between A and $\neg\neg A$ was explained in Section 3.7(c): The former is a direct proposition, the latter expresses the impossibility of something negative, the best example being direct existence against the impossibility of non-existence. The former can be established by showing an object with a required property, the latter by showing that it is impossible that no object has the property.

One reason for the natural tendency to accept the law of double negation, or the related law of excluded middle, is as follows. If there is only a finite number of alternatives, the question of A or $\neg A$ can be decided by going through all of these. Say, if we claim, for natural numbers less than 100, that there are three and only three successive odd numbers that are all prime, we can go through all possible cases and find that 3, 5, and 7 are precisely those three numbers. More generally, the constructive interpretation of $A \vee \neg A$ is that it expresses the **decidability** of A.

The laws of excluded middle and of double negation are characteristic of **classical** logic. Their special nature was first realized in a general way by L. Brouwer in his thesis in 1907. By 1930, classical and intuitionistic logic were clearly delimited in the axiomatic tradition of logic: The former was obtained from the latter by the addition of the law of double negation or of excluded middle. Correspondingly, we can add to the list of axioms I.1–IV.9 of Section 3.6(a) the following:

IV Classical double negation axiom

10. $\neg\neg A \supset A$

(a) The rule of indirect proof. Gentzen's natural deduction covered intuitionistic logic. He tried various ways in which one could obtain a system of natural deduction for classical logic **NK** from the one for intuitionistic logic **NI**, such as the addition of a **rule of double negation** by which one could conclude A from $\neg\neg A$. In the axiomatic logical tradition, the corresponding step had been taken by the addition of the law of double negation. Later, since Prawitz' book of 1965, it has been customary to add a **rule of indirect proof** to the introduction and elimination rules of intuitionistic natural deduction:

Table 5.1 The rule of indirect proof

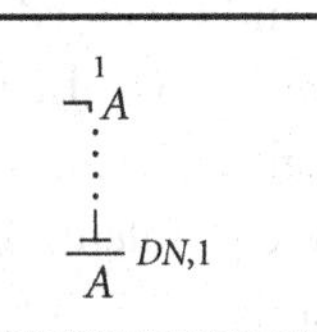

Sometimes the nomenclature *RAA* is used; it stands for *reductio ad absurdum*, the mediæval Latin name of the principle. Given the derivation of the premiss $\bot$ from $\neg A$, rule $\supset I$ would give the conclusion $\neg\neg A$, so that the law of double negation gives the conclusion of rule *DN* when its premiss has been derived. In the other direction, we can derive $\neg\neg A \supset A$ by rule *DN* as follows:

$$\dfrac{\dfrac{\dfrac{\overset{2}{\neg\neg A} \quad \overset{1}{\neg A}}{\bot}\,{\supset}E}{A}\,DN{,}1}{\neg\neg A \supset A}\,{\supset}I{,}2$$

The rule and the axiom have the same deductive strength.

It is quite common to mix genuine indirect proofs with proofs of negative assertions, even in books on logic. A typical example from mathematics concerns proofs of irrationality of a real number: The contrary is assumed, i.e., it is assumed that a given real number c is rational. By deriving a contradiction from this assumption, irrationality of c can be concluded. To see the error in the claim that this is an indirect proof, consider the property *c is an irrational number*. It means, by definition, that there do **not** exist integers n, m such that $c = \frac{n}{m}$. Thus, to prove that c is irrational, assume that there are two such integers. If a contradiction follows, a direct proof of the negative claim that c is irrational has been found, not an indirect one.

If the conclusion in *DN* is a negation $\neg A$, the inference is of the form:

$$\dfrac{\begin{array}{c}\overset{1}{\neg\neg A}\\ \vdots \\ \bot\end{array}}{\neg A}\,DN{,}1$$

If instead of *DN* rule $\supset I$ is applied, the conclusion is $\neg\neg\neg A$. Two negations can be dropped, for as we have seen, $\neg\neg\neg A \supset \neg A$ is a theorem of intuitionistic logic. Then $\neg A$ follows even without indirect inference and we have:

The form of genuine indirect proof in propositional logic. *A genuine indirect proof in propositional logic ends with a positive conclusion.*

The relation between the laws of double negation and excluded middle is somewhat complicated and often misunderstood, sometimes also in the logical literature. One reason is perhaps that if classical reasoning is allowed, both are theorems, and all theorems are trivially equivalent. Therefore, we consider the relation of these laws by the standards of intuitionistic logic: Given $A \vee \neg A$, if $\neg\neg A$ is assumed, A follows, and therefore the implication $\neg\neg A \supset A$ follows from $A \vee \neg A$. An easy proof search in intuitionistic sequent calculus gives a formal derivation for:

$$A \vee \neg A \supset (\neg\neg A \supset A)$$

A similar proof search for the converse fails, for there is no derivation of:

$$(\neg\neg A \supset A) \supset A \vee \neg A$$

Even if this formula is intuitionistically underivable, the law of excluded middle is derivable by the **rule** of indirect proof:

$$\dfrac{\dfrac{\overset{2}{\neg(A \vee \neg A)} \qquad \dfrac{\dfrac{\dfrac{\overset{2}{\neg(A \vee \neg A)} \quad \dfrac{\overset{1}{A}}{A \vee \neg A}\,{\vee I}}{\bot}\,{\supset E}}{\neg A}\,{\supset I,1}}{A \vee \neg A}\,{\vee I}}{\bot}\,{\supset E}}{A \vee \neg A}\,{DN,2}$$

As noted, rule *DN* and the law of double negation are equivalent, and we have another way to see that the law of excluded middle follows from *DN*: We have shown earlier that $\neg\neg(A \vee \neg A)$ is a theorem in intuitionistic logic. Therefore the negation of excluded middle, $\neg(A \vee \neg A)$ gives at once a contradiction and the rule of indirect proof the conclusion $A \vee \neg A$. The overall situation is that if $\vdash \neg\neg A \supset A$, then $\vdash A \vee \neg A$, but **not** $\vdash (\neg\neg A \supset A) \supset (A \vee \neg A)$. More generally:

Let B be derivable whenever A is, i.e., let $\vdash B$ whenever $\vdash A$. It is a logical fallacy to conclude $\vdash A \supset B$.

For another example, consider the disjunction property of intuitionistic logic: If $\vdash A \vee B$ then $\vdash A$ or $\vdash B$. Still, neither of $\vdash A \vee B \supset A$ and $\vdash A \vee B \supset B$ holds. In the proof of the disjunction property, it is assumed that $\vdash A \vee B$. Next it is observed that the last rule must be $R\vee$, by which the disjunction property follows. For the derivability of an implication as in the example $A \vee B \supset A$, a derivation of A from the assumption $A \vee B$ would be required, i.e., a derivation of $A \vee B \vdash A$.

There is thus a difference between *the assumption A* and *the assumption that A is derivable* that can be expressed by:

When we assume A, we assume that there is a hypothetical derivation of A. When we assume $\vdash A$, we assume that there is an actual derivation of A.

The explanation of $A \supset B$ in terms of proof was that by a proof of $A \supset B$, any proof of A can be turned into some proof of B. With $\vdash A$, instead, it is claimed that there is some proof of A.

When A is assumed, it must be possible to compose a derivation of A and any derivation from the assumption A. This matter is discussed further in Section 13.4.

Classical logic is in many ways simpler than intuitionistic logic, because it does not make the distinction between a proposition and its double negation. We show, as further examples of classical derivations, that implication and falsity suffice for defining the rest of the connectives.

Examples of classical derivations:

a. $\vdash A \& B \supset\subset \neg(A \supset \neg B)$. The implication from left to right is derivable in intuitionistic logic, so we do the genuinely classical part of the result:

$$
\dfrac{\dfrac{\dfrac{\overset{3}{\neg(A \supset \neg B)} \qquad \dfrac{\dfrac{\dfrac{\overset{3}{\neg(A \& B)} \qquad \dfrac{\overset{2}{A} \quad \overset{1}{B}}{A \& B}{\scriptstyle \&I}}{\bot}{\scriptstyle \supset E}}{\neg B}{\scriptstyle \supset I,1}}{A \supset \neg B}{\scriptstyle \supset I,2}}{\bot}{\scriptstyle \supset E}}{A \& B}{\scriptstyle DN,3}}{\neg(A \supset \neg B) \supset A \& B}{\scriptstyle \supset I,3}
$$

b. $\vdash A \vee B \supset\subset (\neg A \supset B)$. We show the implication from right to left, the other direction being derivable in intuitionistic logic:

$$\dfrac{\dfrac{\dfrac{\overset{2}{\neg(A \vee B)} \quad \dfrac{\dfrac{\overset{3}{\neg A \supset B} \quad \dfrac{\dfrac{\overset{2}{\neg(A \vee B)} \quad \dfrac{\overset{1}{A}}{A \vee B}{\scriptstyle \vee I}}{\bot}{\scriptstyle \supset E}}{\neg A}{\scriptstyle \supset I,1}}{B}{\scriptstyle \supset E}}{A \vee B}{\scriptstyle \vee I}}{\bot}{\scriptstyle \supset E}}{A \vee B}{\scriptstyle DN,2}}{(\neg A \supset B) \supset A \vee B}{\scriptstyle \supset I,3}$$

If in place of A we have $\neg A$ and if one double negation is deleted, the equivalence becomes $\neg A \vee B \supset\subset (A \supset B)$. Other purely classical laws include $A \vee B \supset\subset \neg(\neg A \,\&\, \neg B)$.

(b) Negation as a primitive notion in classical logic. As with intuitionistic logic in Section 3.5(c), we shall have a look at a formulation of natural deduction for classical logic in which negation is a primitive notion. Negation and any one of &, $\vee$, or $\supset$ can then be used as the connective with which the others are definable in classical logic.

The rules for negation are two, the first the same as in the intuitionistic calculus with primitive negation of Table 3.11:

Table 5.2 Classical rules for negation

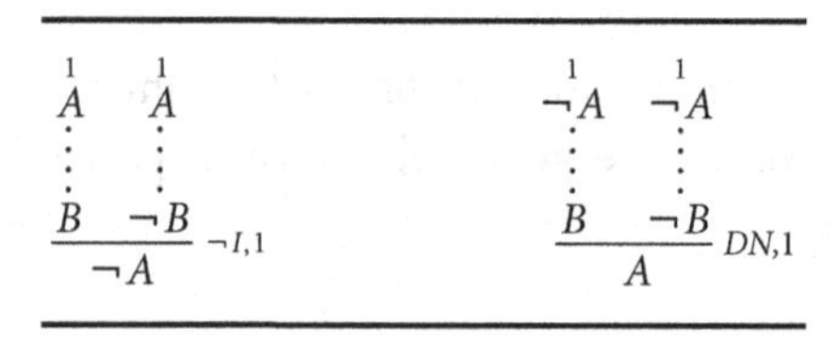

It often happens that the number of closed assumptions is 0 in one of the derivations of the premisses. The second rule is a $\bot$-free variant of the rule of indirect proof of Table 5.1. What corresponds to falsity elimination, or rule $\neg E$ of Table 3.11, is a special case of rule DN when the assumption $\neg A$ has been used 0 times. On the other hand, with B identical to $\neg A$, we have the instance:

$$\dfrac{\overset{1}{\neg A} \quad \neg\neg A}{A}\,{\scriptstyle DN,1}$$

Thus, the rule of double negation elimination becomes a special case of rule *DN* for primitive negation.

5.2 Normal derivations and the subformula property

With rule *DN*, the subformula property can be lost, even if a derivation has none of the convertibilities of intuitionistic natural deduction. It can happen that the conclusion of *DN* is the major premiss of an elimination, as in:

$$\dfrac{\dfrac{\dfrac{\overset{2}{\neg(A \vee \neg A)} \quad \dfrac{\dfrac{\dfrac{\overset{2}{\neg(A \vee \neg A)} \quad \dfrac{\overset{1}{A}}{A \vee \neg A}{\scriptstyle \vee I}}{\bot}{\scriptstyle \supset E}}{\neg A}{\scriptstyle \supset I,1}}{A \vee \neg A}{\scriptstyle \vee I}}{\bot}{\scriptstyle \supset E}}{A \vee \neg A}{\scriptstyle DN,2} \quad \dfrac{\overset{4}{A}}{\neg\neg A \supset A}{\scriptstyle \supset I} \quad \dfrac{\dfrac{\dfrac{\overset{3}{\neg\neg A} \quad \overset{4}{\neg A}}{\bot}{\scriptstyle \supset E}}{A}{\scriptstyle \bot E}}{\neg\neg A \supset A}{\scriptstyle \supset I,3}}{\neg\neg A \supset A}{\scriptstyle \vee E,4}$$

No trace is left of the conclusion $A \vee \neg A$ of *DN*, whereas, had *DN* been followed by an introduction rule, the only thing lost would have been a negation in front of the premiss of the introduction rule. It turns out that the definition of a normal derivation of intuitionistic logic of Section 3.4 is sufficient to guarantee the crucial subformula property also for classical natural deduction:

Definition 5.1. Normal derivation in classical logic. *A derivation in* **NK** *is* **normal** *if all major premisses of E-rules are assumptions.*

The process of normalization contains steps in which instances of *DN* are permuted down whenever their conclusions are major premisses in *E*-rules. The case of $\vee E$ with a major premiss derived by *DN*, as in the above example, is transformed in the following way:

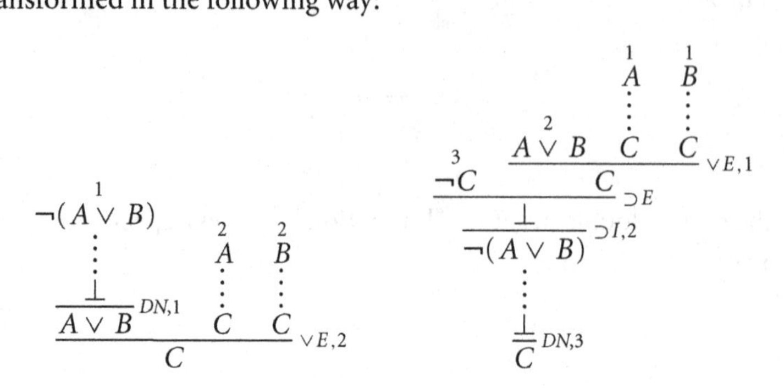

Rule *DN* is permuted down to give the conclusion C of $\vee E$ in the given derivation. These permutations work in the same way for the other elimination rules and are left as exercises. They work also when the premiss of an introduction rule of propositional logic has ben derived by *DN*, the first case being $\& I$ as in:

$$\dfrac{\dfrac{\begin{matrix}\overset{1}{\neg A}\\ \vdots\\ \bot\end{matrix}}{A}{\scriptstyle DN,1} \qquad \begin{matrix}\vdots\\ B\end{matrix}}{A \& B}{\scriptstyle \& I} \qquad\qquad \dfrac{\begin{matrix}\dfrac{\dfrac{\overset{3}{\neg(A \& B)} \quad \dfrac{\overset{1}{A} \quad \begin{matrix}\vdots\\ B\end{matrix}}{A \& B}{\scriptstyle \& I}}{\bot}{\scriptstyle \supset E}}{\neg A}{\scriptstyle \supset I,1}\\ \vdots\\ \bot\end{matrix}}{A \& B}{\scriptstyle DN,2}$$

The second case is $\vee I$:

$$\dfrac{\dfrac{\begin{matrix}\overset{1}{\neg A}\\ \vdots\\ \bot\end{matrix}}{A}{\scriptstyle DN,1}}{A \vee B}{\scriptstyle \vee I} \qquad\qquad \dfrac{\begin{matrix}\dfrac{\dfrac{\overset{2}{\neg(A \vee B)} \quad \dfrac{\overset{1}{A}}{A \vee B}{\scriptstyle \vee I}}{\bot}{\scriptstyle \supset E}}{\neg A}{\scriptstyle \supset I,1}\\ \vdots\\ \bot\end{matrix}}{A \vee B}{\scriptstyle DN,2}$$

Finally, we have $\supset I$:

$$\dfrac{\dfrac{\begin{matrix}\overset{1}{\neg B}, \overset{2}{A}\\ \vdots\\ \bot\end{matrix}}{B}{\scriptstyle DN,1}}{A \supset B}{\scriptstyle \supset I,2} \qquad\qquad \dfrac{\begin{matrix}\dfrac{\dfrac{\overset{2}{\neg(A \supset B)} \quad \dfrac{\overset{1}{B}}{A \supset B}{\scriptstyle \supset I}}{\bot}{\scriptstyle \supset E}}{\neg B}{\scriptstyle \supset I,1}\\ \vdots\\ \bot\end{matrix}}{A \supset B}{\scriptstyle DN,2}$$

It can also happen that, permuting down *DN*, another instance of *DN* is met. Then the conclusion of the first is the premiss of the second, so the latter has to be the formula $\perp$ and we have the situation:

$$\cfrac{\cfrac{\begin{matrix} \overset{2}{\neg A}, \overset{1}{\neg\perp} \\ \vdots \\ \perp \end{matrix}}{\perp}\,DN,1}{A}\,DN,2$$

The first closed assumption $\neg\perp$ is provable, and the derivation is transformed into one that has a single instance of *DN*:

$$\cfrac{\begin{matrix} \overset{2}{\neg A}, \cfrac{\overset{1}{\perp}}{\neg\perp}\supset I,1 \\ \vdots \\ \perp \end{matrix}}{A}\,DN,2$$

Collecting all the permutability results, we have:

Theorem 5.2. Normal form for classical propositional logic. *Derivations in* **NK** *can be so transformed that the major premisses of E-rules are assumptions. Rule DN can be permuted down so that there is at most one instance of rule DN as a last rule.*

Corollary 5.3. Subformula property for normal derivations in NK. *All formulas in a normal derivation of A from the open assumptions Γ in* **NK** *are subformulas or negations of subformulas of A, Γ.*

As observed above, instances of rule *DN* with a negative conclusion can be replaced by intuitionistic steps. In particular, by the normal form, we get:

Corollary 5.4. Glivenko's theorem. *If $\neg A$ is derivable from the assumptions Γ classically, it is derivable intuitionistically.*

The result holds also for the formula $\perp$, and therefore the consistency of intuitionistic propositional logic guarantees also the consistency of classical propositional logic with its indirect inferences.

Assume that A is classically derivable. Then also $\neg\neg A$ is classically derivable, so that by Glivenko's theorem, $\neg\neg A$ is intuitionistically derivable.

The converse to Glivenko's theorem is trivially true: Each intuitionistic derivation is a classical derivation, and especially a derivation of a negation, because intuitionistic natural deduction is a subsystem of classical natural deduction. On the other hand, not every classical derivation is an intuitionistic derivation, so that the class of classical logical laws is larger than the class of intuitionistic logical laws. This greater **deductive strength** of classical logic over intuitionistic logic is only apparent: There is for each formula A a classically equivalent formula A^* such that if A is classically derivable, then A^* is intuitionistically derivable. We can say that classical logic has less **expressive power** because it fails to make a distinction into directly provable and only indirectly provable logical laws. In the case of propositional logic, it is sufficient to take for A^* the double-negated formula $\neg\neg A$.

Notes and exercises to Chapter 5

Classical natural deduction without disjunction and with the rule of indirect proof restricted to atoms was invented by Prawitz (1965). The restriction prevents the use of conclusions of indirect proof as major premisses of elimination rules. The idea on which the present treatment is based, namely the use of the normal form of general elimination rules, has a similar effect. It was first carried through in von Plato and Siders (2012).

1. Give derivations in **NK** for the following:
   a. $((A \supset B) \supset A) \supset A$
   b. $(A \supset B) \vee (B \supset A)$
   c. $(A \supset B \vee C) \supset (A \supset B) \vee (A \supset C)$

2. We showed above how rule DN is permuted down if it has been used for the derivation of a major premiss of rule $\vee E$. Work out the analogous permutations when the latter rule is $\& E$ and $\supset E$.

6 | Proof search in classical logic

6.1 Assumptions and cases

We shall give a sequent calculus for classical logic, originally suggested by Gentzen in 1933. In it, the derivability relation is generalized in a surprising way. The derivability relation in the earlier chapters was written $A_1, \ldots, A_n \vdash B$. Repeated application of rule $L\&$ gives, omitting the parentheses, $A_1 \& \ldots \& A_n \vdash B$. We get $\vdash A_1 \& \ldots \& A_n \supset B$ by rule $R\supset$. On the other hand, by the invertibility of rules $R\supset$ and $L\&$, if $\vdash A_1 \& \ldots \& A_n \supset B$ is a derivable sequent, also $A_1, \ldots, A_n \vdash B$ is. Consider next the sequent in which there is in place of B a disjunction $B_1 \vee \ldots \vee B_m$, namely:

$$\vdash A_1 \& \ldots \& A_n \supset B_1 \vee \ldots \vee B_m$$

By the invertibility of rules $R\supset$ and $L\&$, if this sequent is derivable, also $A_1, \ldots, A_n \vdash B_1 \vee \ldots \vee B_m$ is.

Disjunction in general is used for expressing **cases**, as is clearly seen from the disjunction elimination rule of natural deduction. Thus, the sequent $A_1, \ldots, A_n \vdash B_1 \vee \ldots \vee B_m$ expresses that the disjunction $B_1 \vee \ldots \vee B_m$ gives the cases $B_1, \ldots, B_m$ that are derivable from the assumptions $A_1, \ldots, A_n$. It remains to adopt to sequents the notation for a **list of cases** $B_1, \ldots, B_m$:

Table 6.1 The general form of classical sequents

$A_1, \ldots, A_n \vdash B_1, \ldots, B_m$

The derivability relation has now been generalized from one conclusion under given assumptions into a number of possible cases under given assumptions. Lists of assumptions and cases are denoted by Greek capital letters, as in $\Gamma \vdash \Delta$. Here the assumptions Γ form the **antecedent** of the sequent and the cases Δ its **succedent**. Sequents with more than one formula in the succedent are called **multisuccedent** sequents. Gentzen called them often **symmetric sequents.**

The 'comma at left' of the derivability relation corresponds, by rule $L\&$ and its invertibility, to a conjunctive use of assumptions. The 'comma at right' can be treated in an analogous way, by changing the right rules for disjunction: There is just one rule that gives the conclusion $\Gamma \vdash \Delta, A \lor B$ from the premiss $\Gamma \vdash \Delta, A, B$. It will turn out to be invertible.

The cases $B_1, \ldots, B_m$ in a sequent $A_1, \ldots, A_n \vdash B_1, \ldots, B_m$ are **essentially classical** in the sense that the sequent can be derivable, but none of the m sequents with a single case, i.e., an ordinary conclusion, need be derivable:

$$A_1, \ldots, A_n \vdash B_1 \quad A_1, \ldots, A_n \vdash B_2 \quad \ldots \quad A_1, \ldots, A_n \vdash B_m$$

The same holds especially when the antecedent is empty, as in the law of excluded middle $\vdash A \lor \neg A$ that is derivable: By the invertibility of rule $R\lor$, $\vdash A, \neg A$ is derivable, but neither of $\vdash A$ and $\vdash \neg A$ need be derivable.

6.2 An invertible classical calculus

Negation was treated differently from the other connectives in natural deduction. In the sequent calculus for classical logic, negation can be treated in exactly the same way as the other connectives, i.e., with a left and a right rule. Consider a sequent such as $\Gamma \vdash \Delta, A$. If we add to the list of assumptions $\neg A$, then its contrary A cannot occur as a possible case but must be deleted. Similarly, if in a given sequent $A, \Gamma \vdash \Delta$ we add to the list of possible cases at right $\neg A$, then A on the assumption side must be deleted. Formally, we have the two rules of negation:

Table 6.2 The rules of negation in classical sequent calculus

$$\frac{\Gamma \vdash \Delta, A}{\neg A, \Gamma \vdash \Delta}\, L\neg \qquad \frac{A, \Gamma \vdash \Delta}{\Gamma \vdash \Delta, \neg A}\, R\neg$$

These rules are invertible. We can see it by considering first the conclusion of the left rule, namely $\neg A, \Gamma \vdash \Delta$. If to the succedent A is added, $\neg A$ must be deleted from the antecedent, so that the premiss of rule $L\neg$ is reached. The right rule is seen to be similarly invertible.

The two negation rules contain the essence of classical propositional logic in a sequent calculus formulation:

In classical logic, an assumption A can be turned into a case ¬A and the other way around, and a case A can be turned into an assumption ¬A and the other way around.

For more insight into the matter, consider the classically provable logical law

$$(A \supset B) \supset\subset \neg A \vee B$$

In terms of sequent calculus, $A \supset B$ is provable if and only if the sequent $A \vdash B$ is derivable. On the other hand, $\neg A \vee B$ is provable if and only if the sequent $\vdash \neg A, B$ is derivable. This latter sequent follows from $A \vdash B$ by rule $R\neg$. Thus, the classical equivalence of $A \supset B$ and $\neg A \vee B$ is essentially a consequence of rule $R\neg$ and its invertibility.

The rest of the rules of classical sequent calculus are reformulations of the intuitionistic rules with multisuccedent sequents:

Table 6.3 Rules for &, ∨, and ⊃ in classical sequent calculus

$$\frac{A, B, \Gamma \vdash \Delta}{A \& B, \Gamma \vdash \Delta}{\scriptstyle L\&} \qquad \frac{\Gamma \vdash \Delta, A \quad \Gamma \vdash \Delta, B}{\Gamma \vdash \Delta, A \& B}{\scriptstyle R\&}$$

$$\frac{A, \Gamma \vdash \Delta \quad B, \Gamma \vdash \Delta}{A \vee B, \Gamma \vdash \Delta}{\scriptstyle L\vee} \qquad \frac{\Gamma \vdash \Delta, A, B}{\Gamma \vdash \Delta, A \vee B}{\scriptstyle R\vee}$$

$$\frac{\Gamma \vdash \Delta, A \quad B, \Gamma \vdash \Delta}{A \supset B, \Gamma \vdash \Delta}{\scriptstyle L\supset} \qquad \frac{A, \Gamma \vdash \Delta, B}{\Gamma \vdash \Delta, A \supset B}{\scriptstyle R\supset}$$

The rules for conjunction and disjunction display a symmetry: Like the two rules of negation, they are left–right mirror images of each other. Rule $R\vee$ is just like rule $L\&$ with the antecedent and succedent and ∨ and & interchanged, and similarly for the pair $L\vee$ and $R\&$. Note finally that rule $L\supset$ has no repetition of the principal formula in the antecedent of the first premiss.

Negation is treated as the other connectives. Therefore there is no need for a special formula ⊥ and the only sequents that begin a derivation branch are initial sequents of the form:

Table 6.4 Initial sequents in classical sequent calculus

$$A, \Gamma \vdash \Delta, A$$

Looking at the rules, we notice that each premiss has at least one connective less than the conclusion. Therefore the termination of root-first proof search follows in a very simple way: No derivation branch can contain more steps of inference than the number of connectives in the endsequent.

We shall first give some examples of classical derivations, then show that all of the classical rules are invertible.

Examples of classical derivations:

a. $\vdash A \lor \neg A$

$$\dfrac{\dfrac{A \vdash A}{\vdash A, \neg A}\, R\neg}{\vdash A \lor \neg A}\, R\lor$$

The mirror-like **duality** of & and $\lor$, seen in the rules of Table 6.3, suggests the following derivation:

$$\dfrac{\dfrac{A \vdash A}{A, \neg A \vdash}\, L\neg}{A \,\&\, \neg A \vdash}\, L\&$$

The impossibility of a contradiction such as $A \,\&\, \neg A$ is shown by a derivation that has a sequent with an **empty** succedent. It plays the role of the formula $\bot$ of the intuitionistic calculus.

b. $\vdash \neg\neg A \supset A$

$$\dfrac{\dfrac{\dfrac{A \vdash A}{\vdash A, \neg A}\, R\neg}{\neg\neg A \vdash A}\, L\neg}{\vdash \neg\neg A \supset A}\, R\supset$$

A derivation dual to the two uppermost steps gives the sequent $A \vdash \neg\neg A$ from which the reverse implication $\vdash A \supset \neg\neg A$ follows.

c. $\vdash \neg(\neg A \,\&\, \neg B) \supset A \lor B$

$$\dfrac{\dfrac{\dfrac{\dfrac{\dfrac{A \vdash A, B}{\vdash A, B, \neg A}\, R\neg \qquad \dfrac{B \vdash A, B}{\vdash A, B, \neg B}\, R\neg}{\vdash A, B, \neg A \,\&\, \neg B}\, R\&}{\neg(\neg A \,\&\, \neg B) \vdash A, B}\, L\neg}{\neg(\neg A \,\&\, \neg B) \vdash A \lor B}\, R\lor}{\vdash \neg(\neg A \,\&\, \neg B) \supset A \lor B}\, R\supset$$

This derivation is best read in the root-first direction. The reverse of the implication is derivable already in intuitionistic logic, so that $\neg(\neg A \,\&\, \neg B)$

is equivalent to $A \vee B$ in classical logic. This result can be interpreted as follows:

There is no independent notion of disjunction in classical logic, but it can be defined in terms of conjunction and negation.

Note, however, that conjunction is not more 'privileged' than disjunction, because conjunction can be equally well defined by disjunction and negation, through the equivalence of $A \& B$ and $\neg(\neg A \vee \neg B)$ that is provable in classical logic.

d. $\vdash (A \supset B) \supset \neg A \vee B$

$$\cfrac{\cfrac{\cfrac{\cfrac{A \vdash A \quad B \vdash B}{A,\, A \supset B \vdash B}\,L\supset}{A \supset B \vdash \neg A,\, B}\,L\neg}{A \supset B \vdash \neg A \vee B}\,R\vee}{\vdash (A \supset B) \supset \neg A \vee B}\,R\supset$$

The reverse implication is again derivable is intuitionistic logic, so that implication is definable by disjunction and negation. In fact, negation and any one of &, $\vee$, or $\supset$ can be used to define the remaining two connectives.

The above derivations are remarkably easy to construct, in comparison to those of classical natural deduction of the previous chapter. The explanation lies in the following:

Theorem 6.1. Invertibility of the classical rules. *Whenever a sequent that matches the conclusion of a logical rule is derivable, the corresponding premisses of the logical rule are derivable.*

Proof. The proof of invertibility of $L\&$ and $L\vee$ goes through as before, with an arbitrary context Δ in place of the arbitrary formula C in the proof of the inversion lemma of Section 4.2.

For rule $L\supset$, the base case is that a sequent that matches its conclusion is an initial sequent. If in $A \supset B, \Gamma \vdash \Delta$ the implication $A \supset B$ is not a formula in Δ, the corresponding premisses of $L\supset$, namely $\Gamma \vdash \Delta, A$ and $B, \Gamma \vdash \Delta$, are also initial sequents. Otherwise we have the case that $\Delta \equiv \Delta', A \supset B$ and the initial sequent is $A \supset B, \Gamma \vdash \Delta', A \supset B$. The sequents $\Gamma \vdash \Delta', A \supset B, A$ and $B, \Gamma \vdash \Delta', A \supset B$ that correspond to the premisses of rule $L\supset$ with $A \supset B$ principal are derived by:

$$\cfrac{A, \Gamma \vdash \Delta', B, A}{\Gamma \vdash \Delta', A \supset B, A}\,R\supset \qquad \cfrac{A, B, \Gamma \vdash \Delta', B}{B, \Gamma \vdash \Delta', A \supset B}\,R\supset$$

If $A \supset B, \Gamma \vdash \Delta$ has been derived by $L\supset$ with $A \supset B$ as the principal formula, the premisses are derivable. Otherwise it is derived by $L\supset$ on some other antecedent formula. Apply now the inductive hypothesis to the premisses of that rule, and then the rule, as in the proof of invertibility of Section 4.2.

For a proof of invertibility of the negation rules, consider the situation in which an initial sequent is of the form $\neg A, \Gamma \vdash \Delta, \neg A$, the other cases of initial sequents being handled as above. Then the premiss of $R\neg$, namely $A, \neg A, \Gamma \vdash \Delta$, is derived by:

$$\frac{A, \Gamma \vdash \Delta, A}{A, \neg A, \Gamma \vdash \Delta}\, L\neg$$

With rule $L\neg$, invertibility of an initial sequent is proved similarly by a step of rule $R\neg$. If $\neg A, \Gamma \vdash \Delta$ has been derived by a logical rule and the shown occurrence of $\neg A$ was not principal, proceed as above.

For the right rules, the proofs of invertibility go through in an entirely analogous fashion. QED.

We could go on to show the admissibility of contraction at right, in analogy to the admissibility of contraction at left as in Section 4.5, but the result is of no use because the premisses are always simpler than the conclusion when root-first proof search is applied in classical propositional logic.

If the order of two consecutive rules can be changed, the result of root-first decomposition is the same, a property that is verified by going through all the cases. Consider rules $R\&$ and $L\lor$ as an example:

$$\frac{\dfrac{A, \Gamma \vdash \Delta, C \quad A, \Gamma \vdash \Delta, D}{A, \Gamma \vdash \Delta, C \,\&\, D}\, R\& \qquad \dfrac{B, \Gamma \vdash \Delta, C \quad B, \Gamma \vdash \Delta, D}{B, \Gamma \vdash \Delta, C \,\&\, D}\, R\&}{A \lor B, \Gamma \vdash \Delta, C \,\&\, D}\, L\lor$$

Decomposition of the endsequent in reversed order gives:

$$\frac{\dfrac{A, \Gamma \vdash \Delta, C \quad B, \Gamma \vdash \Delta, C}{A \lor B, \Gamma \vdash \Delta, C}\, L\lor \qquad \dfrac{A, \Gamma \vdash \Delta, D \quad B, \Gamma \vdash \Delta, D}{A \lor B, \Gamma \vdash \Delta, D}\, L\lor}{A \lor B, \Gamma \vdash \Delta, C \,\&\, D}\, R\&$$

Exactly the same four premisses were reached. The rest of the cases of permutability of two consecutive rules goes through similarly and is left as an exercise. The overall result is:

Terminating proof search in classical propositional logic. *Given a sequent $\Gamma \vdash \Delta$, if it is decomposed in any order whatsoever and if all topsequents are initial sequents, $\Gamma \vdash \Delta$ is derivable, otherwise it is underivable.*

As with intuitionistic logic, the termination of proof search is an in-principle property. Application of, say, 50 two-premiss rules in a root-first order can produce a derivation with up to 2^{50} initial sequents. The question therefore is: Given that $\Gamma \vdash \Delta$ is underivable, is there a feasible method for finding at least one topsequent that is not an initial sequent?

Notes and exercises to Chapter 6

The invertible propositional sequent calculus of this chapter was invented by the Finnish logician Oiva Ketonen (1944) who had studied with Gentzen in Göttingen in 1938–9. The calculus became well known through the review Bernays (1945). It was used in Kleene's influential book *Introduction to Metamathematics* (1952) and formed the basis of Beth's 'tableau system' (1955), which for the propositional part is just another way of writing Ketonen's rules.

1. Give derivations in classical sequent calculus for the following:
   a. $((A \supset B) \supset A) \supset A$
   b. $(A \supset B) \vee (B \supset A)$
   c. $(A \supset B \vee C) \supset (A \supset B) \vee (A \supset C)$

2. Show that if the order of two given consecutive rule instances in a derivation in classical sequent calculus are permuted, the same premisses are found.

7 | The semantics of propositional logic

The explanation of the notion of a proposition in Section 1.3 required that such propositions be complete declarative sentences that state a possible state of affairs. The notion of logical truth tries to capture the relation between such sentences and states of affairs, and to be formulated relative to classical and intuitionistic ways of reasoning, respectively. We start with the former because it is simpler, then explain the Kripke semantics of intuitionistic propositional logic. In the final section, a completeness proof is given for classical propositional logic that ties closely together the proof system of Chapter 6 and the standard 'truth-table' semantics of this chapter.

7.1 Logical truth

The semantics of classical propositional logic is based on a notion of absolute truth, whatever that may be. Specifically, each atomic proposition will be either **true** or **false**. The concept of truth in classical propositional logic is built on such an assumption:

Basic assumption about truth. *The truth and falsity of atomic propositions in specific circumstances is determined in itself.*

How this determination takes place, whether truth and falsity can be actually determined and known, etc., are questions from which this notion of truth abstracts away: The different possible states of affairs are represented abstractly so that to each of any given atomic formulas $P_1, \dots, P_n$ is assigned a **truth value**, either the value **true** that is abbreviated as **t** or the value **false** that is abbreviated as **f**. The different assignments themselves are denoted by $v, v_1, v_2, \dots$ and the truth values under a specific assignment, say v, by $v(P) = \mathbf{t}$, $v(Q) = \mathbf{f}$, etc., as the case may be. We collect these things in a definition:

Definition 7.1. *A* **valuation** v *over the atomic formulas* $P_1, \dots, P_n$ *is an assignment of one of the two truth values* $v(P_i) = \mathbf{t}$ *and* $v(P_i) = \mathbf{f}$ *to each of the formulas* P_i.

The truth values assigned to compound formulas under a valuation are determined from the truth values of their immediate components, thus, in the end, from the truth values of the atomic formulas. These **truth conditions** are quite natural for conjunction, disjunction, and negation: A conjunction shall be assigned the value **t** if and only if both of the conjuncts have been assigned the value **t**, a disjunction shall be assigned the value **t** if and only if at least one of the disjuncts has been assigned the value **t**, and a negation shall be assigned the value **t** if and only if the unnegated formula has been assigned the value **f**.

For an implication, it is clear that $v(A \supset B) = \mathbf{f}$ if $v(A) = \mathbf{t}$ and $v(B) = \mathbf{f}$. For the remaining three cases, we can use the classical equivalence of $A \supset B$ and $\neg(A \& \neg B)$: If $v(\neg(A \& \neg B)) = \mathbf{t}$, then $v(A \& \neg B) = \mathbf{f}$ by what was said about the value of a negation. Then either $v(A) = \mathbf{f}$ or $v(\neg B) = \mathbf{f}$. In the latter case, $v(B) = \mathbf{t}$. Thus, an implication $A \supset B$ has the value **t** under a given valuation if either A has the value **f** or B has the value **t**, otherwise $A \supset B$ has the value **f**.

If the false formula $\perp$ is taken into use, it has the value **f** for all valuations.

The falsity of a condition in a conditional sentence will make the truth value of the sentence be **true** in classical logic. Consider the sentence *If Italy is in the polar region, there is snow in Hagalund in the winter.* Under the present geographical circumstances, the condition *Italy is in the polar region* is false. Therefore, under these same circumstances, the said conditional sentence is true. This is certainly a counterintuitive notion of truth: One would normally expect that if an implication $A \supset B$ is true, B has something to do with A, which need not be the case according to the concept of truth in classical propositional logic. If we take into use the equivalence of $A \supset B$ and $\neg A \vee B$ of classical logic, our example sentence becomes: *Either it is not the case that Italy is in the polar region, or there is snow in Hagalund in the winter.* This is a perfectly sensible and clearly true sentence under the present geographical conditions: The left disjunct is true and it need not even be known if there is snow in Hagalund in winter or where such a place can be found.

Let us summarize the ways in which the truth values of compound formulas under a given valuation v are determined from the truth values of the components:

Table 7.1 Truth values of compound formulas

$v(A)$	$v(B)$	$v(A \& B)$	$v(A \vee B)$	$v(A \supset B)$	$v(\neg A)$
t	t	t	t	t	f
t	f	f	t	f	f
f	t	f	t	t	t
f	f	f	f	t	t

If a formula contains just one atomic formula, there are two valuations, one in which the atomic formula is assigned the value **t**, and another in which it is assigned the value **f**. If there is a second atomic formula, it can have the value **t** and the first atomic formula can have two values. If the second formula has the value **f**, the first can likewise have two values. Therefore there are altogether four possible valuations. With three atomic formulas, say P, Q, and R, we have:

Table 7.2 Valuations over three atomic formulas

$v(P)$	$v(Q)$	$v(R)$
t	t	t
t	f	t
f	t	t
f	f	t
t	t	f
t	f	f
f	t	f
f	f	f

The first four lines for P and Q are just like the four lines for A and B in Table 7.1. In the valuations of these lines, R is assigned the value **t**. Next the first four lines are repeated for P and Q, but R is changed to have the value **f**, so that all eight lines are different. If a fourth atom S is added, we have first the eight lines of Table 7.2 with S with the value **t**, then another eight lines with S with the value **f**. Now it is seen that each new atom doubles the number of lines that corresponds to the number of possible valuations. In general:

The number of valuations. *If there are n atomic formulas, there are 2^n different possible valuations.*

It is important to keep in mind the notion of a truth value under a valuation, that is, under some given state of affairs, and truth irrespective of such. The latter corresponds to the classical notion of truth:

Definition 7.2. Logical truth. *If A is assigned the value* **t** *under all valuations, it is* **logically true** *or a* **tautology**.

If a sentence has the value **f** in all valuations, it is **logically false**. In the rest of the cases, A is **contingent**: It depends on the circumstances whether it turns out true or false.

It follows from the exponential growth of the number of valuations that **tautology checking** by truth tables can be done feasibly only if the number of atoms in a formula is low. There are, however, cases in which it can be done fast, but it is unknown if such alternative fast tautology checking methods can be found to replace in all cases the exponentially growing number of valuations.

It is easily seen that the computation of the truth value of a formula A is fast when one valuation is given: The number of steps taken in the computation is the same as the number of connectives. Let us take as an example the formula:

$$(P \,\&\, Q \supset R) \supset (P \supset R) \vee (Q \supset R)$$

Let $v(P) = \mathbf{t}$, $v(Q) = \mathbf{f}$, and $v(R) = \mathbf{t}$. We then have $v(P \,\&\, Q) = \mathbf{f}$ and $v(P \,\&\, Q \supset R) = \mathbf{t}$ for the antecedent of the whole implication. The valuation gives $v(P \supset R) = \mathbf{t}$ and $v(Q \supset R) = \mathbf{t}$ for the disjuncts in the consequent, so that $v((P \supset R) \vee (Q \supset R)) = \mathbf{t}$. Therefore the value of the whole formula is:

$$v((P \,\&\, Q \supset R) \supset (P \supset R) \vee (Q \supset R)) = \mathbf{t}$$

To go through all possible valuations, the latter are tabulated as in Tables 7.1 and 7.2, with all the subformulas of the given formula written out. For example, we have for $(P \supset Q) \vee (Q \supset P)$ the table:

$v(P)$	$v(Q)$	$v(P \supset Q)$	$v(Q \supset P)$	$v((P \supset Q) \vee (Q \supset P))$
t	**t**	**t**	**t**	**t**
t	**f**	**f**	**t**	**t**
f	**t**	**t**	**f**	**t**
f	**f**	**t**	**t**	**t**

Thus, the formula $(P \supset Q) \vee (Q \supset P)$ is a tautology, often called by the name **Dummett's law**. It is at the same time a purely classical logical law: We can show by the methods of Chapter 4 that it is underivable in intuitionistic logic. It is also a classical disjunction neither disjunct of which is classically derivable as seen by the methods of the previous chapter. Consider as an instance of Dummett's law the sentence *The infinity of twin primes implies the continuum hypothesis or the continuum hypothesis implies the infinity of twin primes.* Any mathematician believes that a simple conjecture in number theory cannot have anything to do with the still unresolved great problem of axiomatic set theory. Thus, one's natural reaction to the above claim is: *That can't be true! Those two things are not related in any way to each other.*

It is not always necessary to draw a whole truth table with lots of lines to make a tautology check. Whole classes of formulas are not tautologies, for example, all formulas that do not contain a negation or an implication. There is a systematic method by which one can try to **falsify** a formula: One assumes the formula to have the value **f** and then sees if this leads to an impossibility. As an example, consider the truth value assignment:

$$v((P \,\&\, Q \supset R) \supset (P \supset R) \vee (Q \supset R)) = \mathbf{f}$$

The values of the components of the implication are determined: We have $v(P \,\&\, Q \supset R) = \mathbf{t}$ and $v((P \supset R) \vee (Q \supset R)) = \mathbf{f}$. The latter gives $v(P \supset R) = \mathbf{f}$, so $v(P) = \mathbf{t}$ and $v(R) = \mathbf{f}$, as well as $v(Q \supset R) = \mathbf{f}$, so $v(Q) = \mathbf{t}$. With these, $v(P \,\&\, Q \supset R) = \mathbf{f}$, so the whole implication has the value **t**. Therefore the original truth value assignment was impossible and the formula a tautology.

The previous example is again a purely classical logical law, and in fact a counterintuitive one: Consider as atomic formulas order relations between real numbers, written $a < b$. Such order relations are **transitive**: If $a < b$ and $b < c$, then $a < c$. By what was just shown to be a tautology through a failed attempt at falsification, the following is generally true:

$$(a < b \,\&\, b < c \supset a < c) \supset (a < b \supset a < c) \vee (b < c \supset a < c)$$

Again, one's natural reaction is that both of $a < b$ and $b < c$ are needed to conclude $a < c$, and that neither of $a < b$ or $b < c$ alone is sufficient. There is some truth to this, because, in fact, neither of the following is a tautology:

$$(a < b \,\&\, b < c \supset a < c) \supset (a < b \supset a < c)$$
$$(a < b \,\&\, b < c \supset a < c) \supset (b < c \supset a < c)$$

For a simpler example, take:

P: *I am in my fourth year of study,*
Q: *I pass all exams of this year,*
R: *I graduate this year.*

From:

If I am in my fourth year of study and I pass all exams of this year, then I graduate this year,

something that would be true at many schools, follows that:

If I am in my fourth year of study, then I graduate this year, or if I pass all exams of this year, then I graduate this year.

7.2 The semantics of intuitionistic propositional logic

Consider two atoms P, Q in classical logic. By the semantics given in Section 7.1, there are four different situations, or 'possible worlds' that we denote by w_1, w_2, w_3, w_4:

w_1: P, Q
w_2: $P, \neg Q$
w_3: $\neg P, Q$
w_4: $\neg P, \neg Q$

These worlds are static, but we may have come to any one of them by first knowing or having verified nothing about them, then one of P and $\neg P$, and next one of Q and $\neg Q$:

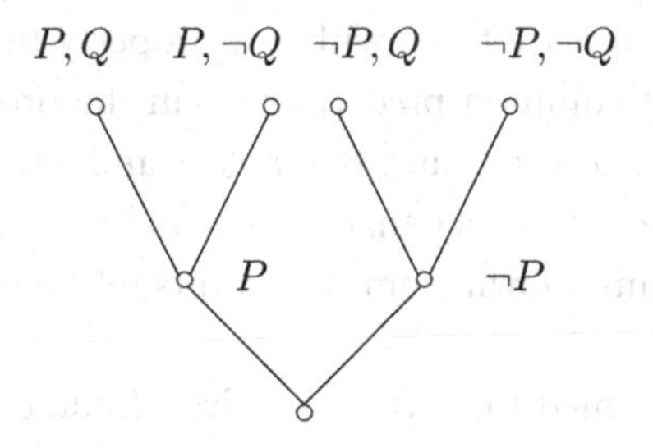

We have first a world in which no truth values are given, then a value is given to P, and next to Q.

Saul Kripke invented in the early 1960s, on the basis of his earlier work on the semantics of **modal logic**, a semantics for intuitionistic logic that makes systematic the idea of a successive verification of truth values of atomic formulas. What are known as **Kripke models** for intuitionistic logic consist of a domain $\mathcal{W}$ of **possible worlds** $w, o, r, \ldots$, a two-place **accessibility relation** $w \leqslant o$ between such worlds, and a **forcing relation** $w \Vdash P$ between worlds and atoms. The last one tells which atoms hold or are 'forced' at what worlds. It is not determined at this stage which negations of atoms are forced in a world.

The accessibility relation comes from modal logic in which the static picture of classical propositional logic does not work well. Consider a situation w in which you hold that P is **possible**, so that you cannot be committed to either of P and $\neg P$. Say, if in some later situation o that is accessible from w you find P to be the case, P must have been possible at w so that $\neg P$ at w is excluded. Similarly, what is possible can turn out later actually not to be the case, so that P at w is likewise excluded.

The accessibility relation is assumed to have the following properties:

Table 7.3 Properties of the accessibility relation

1. There is an **initial world** w_0 such that $w_0 \leqslant w$ for any w in $\mathcal{W}$.
2. The accessibility relation is **reflexive**: $w \leqslant w$ for any w in $\mathcal{W}$.
3. The accessibility relation is **transitive**: If $w \leqslant o$ and $o \leqslant r$, then $w \leqslant r$ for any w, o, r in $\mathcal{W}$.

The accessibility relation is a special case of a **partial order**, a reflexive and transitive two-place relation. (Some pedantics call these 'preorders'.) It is moreover a **tree** with a **root** w_0, with the property that any two worlds w, o have a greatest common predecessor r in the ordering: A common predecessor to w and o is an r such that $r \leqslant w$ and $r \leqslant o$. It is the greatest such predecessor if for any l such that $l \leqslant w$ and $l \leqslant o$, there obtains $l \leqslant r$. Intuitively, if you climb down from two points in the tree, you come to a first common point.

Each world is assumed to force some given finite collection of atoms, possibly none. The forcing relation for formulas with logical structure is defined inductively:

Table 7.4 Definition of the forcing relation

1. $w \Vdash A \,\&\, B$ whenever $w \Vdash A$ and $w \Vdash B$.
2. $w \Vdash A \vee B$ whenever $w \Vdash A$ or $w \Vdash B$.
3. $w \Vdash A \supset B$ whenever from $w \leqslant o$ and $o \Vdash A$ follows $o \Vdash B$.
4. $w \Vdash \neg A$ whenever from $w \leqslant o$ and $o \Vdash A$ follows $o \Vdash C$ for any C.
5. $w \Vdash C$ for any C whenever $w \Vdash A$ and $w \Vdash \neg A$ for some A.

This definition will work for intuitionistic logic with a primitive notion of negation. With a defined notion of negation, clause 5 is often put as: no world forces $\bot$, and clause 4 is left out. It then happens that proofs of the properties of the forcing relation have to rely on somewhat awkward meta-level reasonings that use classical logic. For example, to prove clause 4, one reasons: By 3, $w \Vdash A \supset \bot$ whenever from $w \leqslant o$ and $o \Vdash A$ follows $o \Vdash \bot$. The latter is not the case, so that with $w \leqslant o$, $o \Vdash A$ is impossible for any o. For $w \Vdash \bot \supset C$, one needs: From $w \leqslant o$ and $o \Vdash \bot$ follows $o \Vdash C$. This is the case because $o \Vdash \bot$ is false.

We have chosen, after a suggestion by Sara Negri and Alberto Naibo, the above definition by which the forcing relation for a world w becomes **trivial**, in the sense that it forces all formulas, whenever $w \Vdash A$ and $w \Vdash \neg A$ for some A. We shall from now on pose the requirement:

Nontriviality. *No world must force all formulas.*

To finish the definition of a **Kripke model**, we add the requirement:

Monotonicity. *If $w \Vdash P$ and $w \leqslant o$, then $o \Vdash P$.*

This property holds for arbitrary formulas A:

Lemma 7.3. Monotonicity for arbitrary formulas. *If $w \Vdash A$ and $w \leqslant o$, then $o \Vdash A$.*

Proof. The proof is by induction on the length of formula A:

If A is atomic, the property holds by monotonicity.

If $A \equiv B \,\&\, C$ and $w \Vdash B \,\&\, C$, we have $w \Vdash B$ and $w \Vdash C$ by definition and apply monotonicity to shorter formulas to conclude $o \Vdash B$ and $o \Vdash C$. Now $o \Vdash B \,\&\, C$ follows by definition.

If $A \equiv B \vee C$, the proof goes through similarly.

If $A \equiv B \supset C$, let $w \leqslant o$. To show $o \Vdash B \supset C$, assume $o \leqslant r$ and $r \Vdash B$. By transitivity, $w \leqslant r$ so $w \Vdash B \supset C$ and $r \Vdash B$ give $r \Vdash C$. Therefore $o \Vdash B \supset C$.

If $A \equiv \neg B$, let $w \Vdash \neg B$ and show that if $w \leqslant o$, then $o \Vdash \neg B$. For the latter, assume therefore $o \leqslant r$ and $r \Vdash B$ and show $r \Vdash C$. Transitivity gives $w \leqslant r$, so by $w \Vdash \neg B$, from $r \Vdash B$ follows $r \Vdash C$. Therefore by definition $o \Vdash \neg B$. QED.

Intuitively, whatever holds in a given world, continues to hold in all worlds accessible from that world.

Clause 4 gives as a special case, when A is atomic:

If no world accessible from w forces an atom P, then w forces $\neg P$.

This property is a bit strange: The original idea of Kripke was that worlds are unfolded in time, by finding out new atomic facts. The reasoning about a Kripke model, instead, takes it as a finished structure in which the forcing of formulas in a world w can be decided on the basis of what happens in all the possible future courses of events that start from w. From one point of view, the future is full of possibilities that can become realized, from another point of view, this realization just cuts off alternative histories.

Definition 7.4. Intuitionistic validity. *A formula A is* **valid** *in intuitionistic propositional logic, notation $\Vdash A$, if it is forced in an arbitrary world.*

It is not altogether practical to use this notion for actually proving validity, by considering an arbitrary Kripke model and an arbitrary world in it. Here are some examples:

Example proofs of intuitionistic validity:

a. $\Vdash A \supset (B \supset A \& B)$

Let w be arbitrary and assume $w \leqslant o$ and $o \Vdash A$. For $o \Vdash B \supset A \& B$, assume $o \leqslant r$ and $r \Vdash B$. By monotonicity, $r \Vdash A$, so by definition, $r \Vdash A \& B$. Therefore $o \Vdash B \supset A \& B$, and finally $w \Vdash A \supset (B \supset A \& B)$.

b. $\Vdash A \& B \supset A$

Let w be arbitrary and assume $w \leqslant o$ and $o \Vdash A \& B$. By definition, $o \Vdash A$. Therefore $w \Vdash A \& B \supset A$.

c. $\Vdash (A \supset C) \supset ((B \supset C) \supset (A \vee B \supset C))$

Let w be arbitrary and assume $w \leqslant o$ and $o \Vdash A \supset C$. The task is to show $o \Vdash (B \supset C) \supset (A \vee B \supset C)$, so assume $o \leqslant r$ and $r \Vdash B \supset C$. To show $r \Vdash A \vee B \supset C$, assume $r \leqslant l$ and $l \Vdash A \vee B$. Then $l \Vdash A$ or $l \Vdash B$ by definition. In the first case, we have to show $l \Vdash C$. We get $o \leqslant l$ by transitivity, so $o \Vdash A \supset C$ gives $l \Vdash C$. Then by definition $r \Vdash A \vee B \supset C$, and consequently $o \Vdash (B \supset C) \supset (A \vee B \supset C)$ and finally the sought result $w \Vdash (A \vee B \supset C) \supset (A \supset C) \& (B \supset C)$ for an arbitrary w. The proof in the second case, $l \Vdash B$, is similar.

It is seen that intuitive reasoning by the clauses that define forcing for compound formulas goes parallel to purely syntactic steps of proof in natural deduction. Conjunction introduction corresponds to the first clause of Table 7.4 from right to left and conjunction elimination to the other direction. Disjunction introduction corresponds to the second clause from left to right, and the other direction has the two cases that correspond to disjunction elimination. The clause for implication from right to left lets us conclude $w \Vdash A \supset B$ if $o \Vdash B$ follows from $o \Vdash A$ for any o such that $w \leqslant o$. In the other direction, if $w \Vdash A \supset B$, we get in particular by $w \leqslant w$ that $w \Vdash B$ follows from $w \Vdash A$, i.e., that the forcing relation respects implication elimination.

In the light of the above, it is not surprising that intuitionistic propositional logic turns out to be **sound** and **complete** in relation to Kripke semantics: All theorems are valid and all valid formulas are theorems. We shall prove the former here, the most straightforward way being to show that all the axioms of intuitionistic propositional logic are valid and that rule $\supset E$ preserves validity. The axioms are found in Section 3.6(a) and some were shown valid in the above three examples. That rule $\supset E$ preserves validity was also noted. Therefore axiomatic intuitionistic logic is sound. To prove the same for natural deduction, the notion of validity has to be extended to **validity under assumptions:**

Definition 7.5. Validity under assumptions. *The formula A is* **valid under assumptions** Γ *if, whenever a world w forces all formulas in Γ, also $w \Vdash A$.*

Theorem 7.6. Soundness of intuitionistic natural deduction. *If A is derivable from the assumptions Γ, then A is valid under the assumptions Γ.*

Proof. The proof is by induction on the last rule applied in a derivation. In the case of an assumption A, $w \Vdash A$ gives $w \Vdash A$, so A is valid under the assumption A. If the last rule is $\& I$, assume the premisses A and B

valid under the assumptions Γ and Δ, respectively. Then $w \Vdash A \& B$ is by definition valid under the assumptions Γ, Δ. All other rules in which assumptions are not closed are seen to preserve validity under assumptions in the same way. For rule $\supset I$, assume B to be valid under the assumptions A, Γ, i.e., assume $w \Vdash B$ whenever $w \Vdash A$ and w forces all the formulas in Γ. To show that $w \Vdash A \supset B$ when w forces all the formulas in Γ, let $w \leqslant o$ and $o \Vdash A$. By monotonicity, o forces Γ, so by the inductive hypothesis, $o \Vdash B$. Therefore $w \Vdash A \supset B$, and $A \supset B$ is valid under the assumptions Γ. The case of rule $\vee E$ is handled similarly. QED.

The notion of validity under assumptions can be expressed also in terms of sequents, by considering the above definition to define the validity of a sequent $\Gamma \vdash A$.

Kripke models are more interesting as a method for disproving intuitionistic validity than proving it. The method consists in finding a **countermodel**, i.e., a world that does not force a formula. It is customary to depict such models directly as trees, such as:

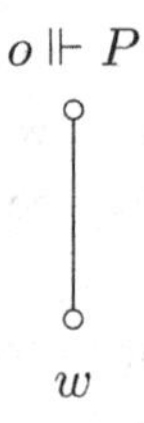

No atom is forced in w. Let's check if $P \vee \neg P$ is forced in w. If so, one of the disjuncts has to be forced and it is not P. If it is $\neg P$, we have $o \Vdash \neg P$ by monotonicity, and $o \Vdash C$ for any C so that o is trivial. Therefore $w \nVdash \neg P$, so the law of excluded middle has an instance that is not intuitionistically valid, and neither is the law for an arbitrary formula in place of P.

Consider next Dummett's law instantiated for two atomic formulas, $(P \supset Q) \vee (Q \supset P)$. Let the Kripke model be:

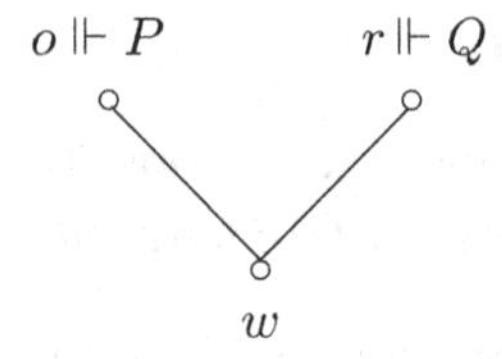

Let us check if $w \Vdash (P \supset Q) \vee (Q \supset P)$. By definition, the first case is $w \Vdash P \supset Q$, so we need to have $l \Vdash Q$ whenever $w \leqslant l$ and $l \Vdash P$. We do have $w \leqslant o$ and $o \Vdash P$, but $o \Vdash Q$ fails. Similarly, w cannot force $Q \supset P$, therefore neither $(P \supset Q) \vee (Q \supset P)$.

Let us take as a final example the classical tautology:

$$(P \,\&\, Q \supset R) \supset (P \supset R) \vee (Q \supset R)$$

How to construct a Kripke countermodel for it? We should have that each time we have moved from the initial world so that both P and Q have become forced, also R should be forced, whereas if only one of them has been forced, a world remains accessible in which $\neg R$ gets forced:

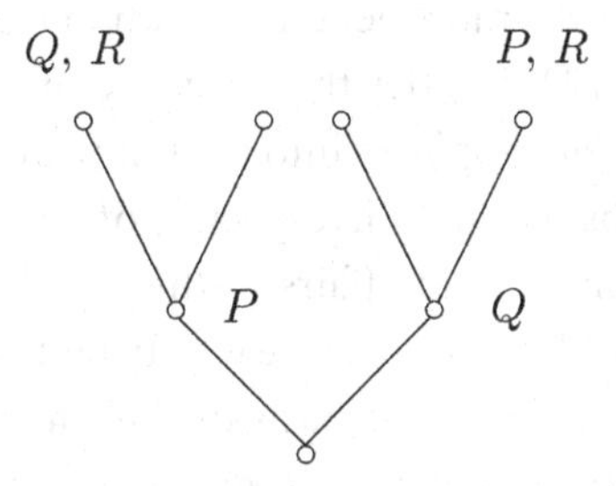

There are two terminal worlds highest up that force $\neg R$. Therefore, the forcing of P does not lead to the forcing of R, and similarly for Q, but whenever both are forced, also R is.

We started the study of logical reasoning with rules that led to intuitionistic logic. Classical logic came out as a special case in which more is assumed, namely the applicability of the law of excluded middle or the principle of indirect proof. It is the same with semantics, even if we described first the semantics of classical propositional logic: The latter comes out as a special case of Kripke semantics, when there is just one single possible world w. The forcing relation determines which atoms hold at w and as we have seen, since there are no new accessible worlds, it happens that if an atom P is not forced in w, its negation is. We have therefore what corresponds to a single valuation, or line in a truth table.

Let us finish with a verification of the law of excluded middle in the case of a single possible world: All Kripke models have the structure:

$$w \Vdash P, \ldots, Q$$

$$\circ$$

We have pointed out that if an atom P is not forced at any world accessible from w, its negation is forced in w. It follows that either $w \Vdash P$ or $w \Vdash \neg P$ for any atom P, consequently $w \Vdash P \vee \neg P$, for an arbitrary w. Therefore $\Vdash P \vee \neg P$, and from this follows next $\Vdash A \vee \neg A$ for arbitrary formulas. The result is left as an exercise.

7.3 Empty tautologies?

Logical truth, of which we saw the definition for the case of propositional logic in Section 7.1, was the basic notion of a philosophy known as **logicism**. Its pioneers were Frege and Russell. The former had invented classical predicate logic in 1879, and both had in turn tried to show that all of mathematics reduces in some way to logic. The concept of logical truth was the cornerstone of the reduction; its specific aim was to show that mathematical truth is a special case of the more general notion of logical truth. This latter notion evolved in the 1920s under the influence of Ludwig Wittgenstein's book *Tractatus Logico-Philosophicus* into the following doctrine: The truths of logic are those truths that hold irrespective of any circumstances. Thus, they do not exclude any state of affairs in the world, therefore do not say anything about the world. They are mere empty tautologies, devoid of content and always true purely on the basis of their form. Truth is found by analysing the form of the sentences of logic until it is seen that the form is empty.

In the next step, following the logicist thesis, mathematical truth was to be shown a special case of logical truth. Frege tried to do this mainly for the case of the concept of a natural number and arithmetic more generally. A general attempt was made by Russell in cooperation with Alfred Whitehead in a three-volume massive undertaking, *Principia Mathematica*, published in 1910–13. There were difficulties in the formal carrying through of the logicist programme, and it has been considered a failure for a long time. That was not, however, the situation in the 1920s.

What does one know when one knows a mathematical truth? The answer in the 1920s was that one knows nothing, because such truths are mere tautologies. One should have thought that if that was the answer, then there must have been something wrong in the assumptions that led to it, but no, the empty theory of truth survived and even prospered, the reason perhaps being the absence of viable alternatives. Eighty years of development in foundational studies has provided some such.

First of all, knowledge of a mathematical truth comes from a proof, so one knows a proof. It then depends on the proof what it is good for. Typical truths in logic and mathematics are in the form of an implication $A \supset B$. If the implication has been proved constructively, i.e., without the use of the principles of classical logic, it provides an **algorithm** and the following is the case:

By a constructive proof of $A \supset B$, any constructive proof of A can be converted into some constructive proof of B.

One of the common forms of a theorem in mathematics is that there is a number of assumptions $A_1, \ldots, A_n$ with some numerical content from which follows a consequence B likewise with some numerical content. The assumptions would be typically expressed with variables, as in: *Let x_1 be a number such that $A_1, \ldots$, let x_n be a number such that A_n*, and the conclusion could be: *y is a number such that B.* A proof of $A_1, \ldots, A_n$ consists in finding some specific values a_i for the x_i and in computations that verify for these values the properties expressed by $A_1, \ldots, A_n$. The proof of B from $A_1, \ldots, A_n$ converts these numerical verifications, i.e., a constructive proof as required above, into a number b in place of y and a verification that it has the property expressed by B. The algorithm that makes the conversion can be made into a computer program. When the data about $A_1, \ldots, A_n$ are fed in and the command to execute the program given, something very specific is going to happen: An electromagnetic program-guided physical process in the computer produces figures with numerical content on the screen. If they were, say, about where a space shuttle is going to be found at a given time when the numerical data concerned its launch and where it was headed, the shuttle will be found there. If the data were about the positions of the planets and the program computed solar eclipses backward a couple of thousand years, some of these eclipses can be found in the records of the ancient astronomers, with their day and duration corresponding to the figures produced on our computer screen. Knowledge of a mathematical truth can thus be the possession of something that can tell how parts of the world work.

One might say that the above discussion is about something that can be described as a process, as a step-by-step proof can, rather than a claim with a direct content. A similar distinction is found in geometry: Part of geometry consists in the proving of theorems, part in the solving of geometric problems. What is the order of conceptual priority here? This question was debated already in ancient times, but there is a Solomonic solution by which they are two aspects of the same thing: A theorem can be seen as a problem to construct a proof, and a problem as a theorem stating that there is a solution to the problem.

The above discussions boil on the question of truth and proof. What is the order of conceptual priority here? A proof expresses knowledge and one can construct the notion of truth on the possibility, in principle, of giving a proof. This was the natural order in which we proceeded in the first six chapters. On the other hand, one can think that truth is the basic thing, absolute and determined in itself, and that knowledge is just a marginal note to truth. Truth is not affected by our knowledge of it. The deep question, then, is: Can there be truths that are unknowable in principle? The one who constructs truth on proof is the philosophical idealist who says that there is no truth outside the world we have constructed. The philosophical realist, instead, says about unknowable truth that it is possible, or that at least we have so far not found any argument to the contrary. Elementary logic is a field in which some of these basic questions about knowledge can be confronted in a very precisely defined setting.

Elementary logic is even epistemology in laboratory conditions.

7.4 The completeness of classical propositional logic

It is time to turn to more earthly matters, namely the relation of derivability to the notion of logical truth in classical propositional logic. We shall show simultaneously that all theorems are tautologies, by which the sequent calculus for classical propositional logic of Chapter 6 is **sound**, and that all tautologies are theorems, by which it is **complete**.

The following prescription will produce what is called the **conjunctive normal form** of a formula. Let the formula be A, and decompose root first the sequent $\vdash A$ by the rules of classical sequent calculus until there are no connectives left anywhere. Each topsequent will have the form:

$$P_1, \ldots, P_m \vdash Q_1, \ldots, Q_n \tag{1}$$

Now move the P_i to the right by rule $\neg R$, then replace the commas at right by disjunctions through rule $\vee R$. It will be convenient not to write out the parentheses. For uniqueness, one can agree that a formula with parentheses left out, as in $P \vee Q \vee R \vee S$, is an abbreviation for $((P \vee Q) \vee R) \vee S$, etc. Now the form of the sequents (1) has become:

$$\vdash \neg P_1 \vee \ldots \vee \neg P_m \vee Q_1 \vee \ldots \vee Q_n \tag{2}$$

Let the sequents produced in this way be $\vdash A_1 \ \ldots \ \vdash A_k$. Now apply rule $R\&$, with a convention on parentheses analogous to that for disjunction, to form:

$$\vdash A_1 \,\&\, \dots \,\&\, A_k \tag{3}$$

Formula A is derivable if and only if the topsequents (1) are initial sequents and the same for formula $A_1 \,\&\, \dots \,\&\, A_k$. It is clear by the construction that a formula is classically equivalent to its conjunctive normal form:

$$\vdash A \supset\subset A_1 \,\&\, \dots \,\&\, A_k \tag{4}$$

To prove that the sequent calculus for classical propositional logic is sound and complete, we need to show that the property of being an initial sequent is maintained if the decomposition of a sequent of the form $B, \Gamma \vdash \Delta, B$ is continued until no connectives remain:

Lemma 7.7. *If a sequent of the form $B, \Gamma \vdash \Delta, B$ is decomposed root first until no connectives remain, only initial sequents are reached.*

Proof. Take any decomposition as required by the lemma. There are five cases according to the form of B:

If B is an atom, the decomposition of formulas in Γ and Δ will keep one copy of B in the antecedent and another in the succedent, so that when no connectives are left, initial sequents of the form $B, \Gamma' \vdash \Delta', B$ are reached, with Γ', Δ' consisting of only atoms.

If B is a compound formula, the decomposition can be done as when B was an atom, so that in the end B is the only compound formula in the sequents of the form $B, \Gamma' \vdash \Delta', B$. There are now four cases:

B is a conjunction $C \,\&\, D$. It is decomposed as in:

$$\cfrac{\cfrac{C, D, \Gamma' \vdash \Delta', C \qquad C, D, \Gamma' \vdash \Delta', D}{C, D, \Gamma' \vdash \Delta', C \,\&\, D}\;R\&}{C \,\&\, D, \Gamma' \vdash \Delta', C \,\&\, D}\;L\&$$

If B is a disjunction, the procedure is dual to the above.

If B is an implication $C \supset D$, the decomposition is:

$$\cfrac{\cfrac{C, \Gamma' \vdash \Delta', C \qquad D, \Gamma' \vdash \Delta', D}{C \supset D, C, \Gamma' \vdash \Delta', D}\;L\&}{C \supset D, \Gamma' \vdash \Delta', C \supset D}\;R{\supset}$$

Finally, if B is a negation $\neg C$, decomposition by rules $\neg L$ and $\neg R$ will turn $\neg C, \Gamma' \vdash \Delta', \neg C$ into $C, \Gamma' \vdash \Delta', C$.

Repetition of the above decompositions will in the end give initial sequents in which all formulas are atomic. QED.

Theorem 7.8. Soundness and completeness of classical propositional logic. *If a formula is derivable in the sequent calculus for classical propositional logic, it is a tautology. If a formula is a tautology, it is derivable in the sequent calculus for classical propositional logic.*

Proof. Let $\vdash A$. Decompose root first until initial sequents are reached. By the lemma, this can be continued until sequents with only atomic formulas are reached. Each of them has at least one identical atom P in the antecedent and succedent. When the sequent is turned into one of the form (2), a part of it is, after possible change of order of the disjuncts, $P \vee \neg P$. Therefore all the formulas $A_1, \ldots, A_k$ in (3) are tautologies, and therefore formula A itself is a tautology.

Assume, for the proof of completeness, that A is a tautology. Form its conjunctive normal form $A_1 \,\&\, \ldots \,\&\, A_k$ as described above. Each conjunct is a tautology of the disjunctive form $\neg P_1 \vee \ldots \vee \neg P_m \vee Q_1 \vee \ldots \vee Q_n$. This can be the case only if some P_i is identical to some Q_j, but then decomposition will result in initial sequents, by which A is derivable. QED.

Notes and exercises to Chapter 7

The truth value semantics of classical propositional logic is present already in the work of Boole of about 1850, to be described in Section 14.2(a). The same holds for Frege's work, cf. Section 14.3(a). The semantics for intuitionistic propositional logic was presented in Kripke (1965).

1. Show that forcing emulates the structural rules of sequent calculus:
   a. If $w \Vdash B$, then $w \Vdash A \supset B$.
   b. If $w \Vdash A \supset (A \supset B)$, then $w \Vdash A \supset B$.
   c. If $w \Vdash A \supset B$ and $w \Vdash B \supset C$, then $w \Vdash A \supset C$.

2. Show that **Peirce's law**, $((A \supset B) \supset B) \supset B$, is strictly classical, by a Kripke model that refutes it.

PART II

Logical reasoning with the quantifiers

8 | The quantifiers

8.1 The grammar of predicate logic

Predicate logic starts from propositional logic and adds to it two things: The atomic formulas receive an inner structure, and **quantifiers** are added, one for expressing **generality**, another for expressing **existence**.

The structure of the atomic formulas is as follows: We have some given collection of individuals, denoted as a whole by $\mathcal{D}$ and called the **domain**, and individuals in the domain, called **objects** and denoted by $a, b, c, \ldots$, $a_1, a_2, \ldots$ etc. Each atomic formula gives a **property** of the objects in $\mathcal{D}$, or a **relation** between several such objects. The notation and reading of atomic formulas is exemplified by the following:

$P(a)$, *object a has the property* $P(a)$
$Q(a, b)$, *objects a and b stand in the relation* $Q(a, b)$ *to each other*
$R(a, b, c)$, *objects a, b, and c stand in the relation* $R(a, b, c)$ *to each other*

For a concrete example, let $\mathcal{D}$ consist of the natural numbers $0, 1, 2, \ldots$, and let P be the property *to be a prime number*. We can form atomic formulas by writing numbers in the argument place of P, say $P(17)$ that is the proposition *17 is a prime number*. Let Q be the order relation $<$ between two natural numbers. Then $Q(7, 5)$ is the proposition *7 is smaller than 5*.

Let, for another example, $\mathcal{D}$ consist of geographical locations found in some given atlas. Consider the relation *location a is between locations b and c*. The relation would be clear when, say, one railway station is between two others, or one location is to the north, the other to the south of a third location, but how much deviation can there be? We need not answer such possible questions, or define what relations in general are. It is sufficient for the purposes of logic that there are clear-cut examples of relations and a clear understanding of what it means for two given objects to stand in such a clear-cut relation to each other.

When the structure of natural language is analysed with predicate logic, there is some freedom in the choice of predicates and relations. It depends on the context of a sentence what relations are meant by the user of a language.

Consider the sentence *Helsingfors is by the Baltic Sea.* The discourse could be about aspects of the Baltic Sea, or aspects of Helsingfors, or more specifically about capitals of the countries that share coastline of the Baltic Sea, etc. If we fix one such relation, say x *lies by* y, we can substitute *Helsingfors* for x to get *Helsingfors lies by* y. This can be considered a one-place predicate. Similarly, for each constant that can be substituted for y, a one-place predicate is obtained, as in x *lies by the Thames.*

The relativity of the choice of predicates is connected to the relativity of the choice of a domain. A domain can be made sufficiently large to accommodate all types of objects encountered within a discourse. Then, within the domain, types of objects can be identified by predicates. Say, we can take as the domain the real numbers, then consider the rationals as reals with a certain property, the natural numbers as reals with a further property, and so on.

There is clearly a limit to how many argument places there can be in a relation that can appear in a natural discourse. It is certainly at least four: If the domain consist of persons, the atomic formula *John's feelings towards his father are like those of Oedipus' to his father* expresses a **four-place** relation. In general, an n-place relation is given by an atomic formula $S(a_1, \ldots, a_n)$. Such relations can have uses for any number n. For an example, consider the coordinates of a point in an n-dimensional space and the relation: x *is the* i*-th coordinate of the point* $(x_1, \ldots, x_n)$.

As the terminology of arguments suggests, properties and relations can be represented as functions. This idea was one of the central ones when Frege presented in 1879 the language of predicate logic, in his little book *Begriffsschrift: Eine nach der arithmetischen nachgebildete Formelsprache des reinen Denkens* (advanced learners who want to refer to the original should notice and keep in mind the double-s in the title). The translation of Frege's book title is something like 'Conceptual notation: A formula language for pure thought, built upon the model of arithmetic.' Frege wanted to analyse sentences into **function** and **argument**, instead of the analysis into subject and predicate of traditional school grammar.

Frege's idea can be made precise through a categorial grammar for predicate logic, in which atomic formulas are values of functions. The notation of functional application, as in Section 1.6, shows how the thing goes:

Table 8.1 A one-place predicate

$$\frac{P : \mathcal{D} \rightarrow \mathcal{F} \quad a : \mathcal{D}}{P(a) : \mathcal{F}}$$

We have a function P from $\mathcal{D}$ to $\mathcal{F}$, the latter the category of formulas, and an argument a in $\mathcal{D}$. The function value $P(a)$ is in $\mathcal{F}$. If we have two arguments, the notation is:

Table 8.2 A two-place predicate

$$\frac{Q : \mathcal{D} \times \mathcal{D} \rightarrow \mathcal{F} \quad a : \mathcal{D} \quad b : \mathcal{D}}{Q(a, b) : \mathcal{F}}$$

The notation $\mathcal{D} \times \mathcal{D}$ indicates that the function takes two arguments, which can be generalized to any number of arguments. What is sometimes called the 'arity of a predicate', i.e., the n of an n-place relation, is usually read from the notation, and we have **unary, binary, n-ary**, etc. predicates.

Next to the **constants** $a, b, c \ldots$ that name objects in a domain $\mathcal{D}$, we need **variables** $x, y, z, \ldots, x_1, x_2, \ldots$ for the expression of **generality** and **existence**. The **universal quantifier** is used for the former, the **existential quantifier** for the latter. The notation follows the idea of suitably inverted letters, here $\forall$ for *All* and $\exists$ for *Exists*. Below, a one-place atomic formula has been quantified and possible readings of the formulas given:

Table 8.3 The quantifiers and some of their readings

$\forall x P(x)$:	*for every x*, $P(x)$,	*all x have the property* $P(x)$
$\exists x P(x)$:	*there is an x such that* $P(x)$,	$P(x)$ *for some x*

We say that the variable x occurs **quantified** in these formulas. If one asks what variables are, the answer is that they are symbols used according to certain rules. Constants, instead, are expressions that name specific objects. There is no handy way of expressing generality and existence without the use of variables. We could say in a single case something like 'all objects have the property P', and 'there is an object with the property P'. However, when formulas grow complex, a way is needed for keeping track of which quantifier quantifies what variable occurrences. Thus, we write in general:

$$\forall x(A) \quad \text{and} \quad \exists x(A)$$

Now it is clear that formula A is quantified. It forms the **scope** of the quantifiers $\forall x$ and $\exists x$.

Definition 8.1. Free and bound variables. *An occurrence of a variable x in a formula A is* **free** *if it is not in the scope of any quantifier* $\forall x$ *or* $\exists x$*. Otherwise the occurrence is* **bound.**

To cut down the number of parentheses, we leave them out around atomic formulas. We leave them out also if a quantifier is followed by another quantifier, as in:

$$\forall x \exists y(P(x, y) \supset Q(x, y, z))$$

The scope of $\exists y$ is the formula in parentheses after it, and the scope of $\forall x$ is what is left if $\forall x$ is deleted. We can read the formula as: *There exists for every x a y such that if $P(x, y)$, then $Q(x, y, z)$.* The formula has one two-place predicate and one three-place predicate, and two bound occurrences of each of x and y, and one free occurrence of z. (The x in $\forall x$ and y in $\exists y$ are not counted.) We shall soon see what the point and meaning of such free occurrences of variables is.

Predicate logic is a language that is used for expressing properties of given individuals and relations among these individuals. One can say that it contains a kind of ontology of the world, namely one of individuals and their properties and relations.

Predicate logic puts no limit to how complicated quantificational structures can be. A sequence of quantifiers such as $\forall x \exists y \forall z$ is quite natural. *There is for every person some problem on which every attempt will fail.* The three-place relation here is: *person x fails on problem y in attempt z.* This example also illustrates the choice of a domain. In this case the domain seems to contain, at least, persons, problems, and attempts. There is a three-place relation that we can write as $Fail(x, y, z)$, in which x is a person, y a problem, and z an attempt. These have to go in the right places in the relation. The usual way to do it is to introduce predicates such as $Person(x)$, $Problem(y)$, and $Attempt(z)$, and to write the formalization as in:

$$Person(x) \,\&\, Problem(y) \,\&\, Attempt(z) \supset Fail(x, y, z)$$

The condition for $Fail(x, y, z)$ can be lifted only if x, y, z are appropriately chosen.

For another, real-life, example, take *There is for every convergent sequence $x_1, x_2, \ldots$ of real numbers a number n such that for every m, if $m > n$, then the difference between x_m and x_{m+1} is less than $1/n$.*

Thinking of the quantifiers, some logical connections are obvious, as in the following:

$$\forall x A \supset \neg \exists x \neg A$$
$$\exists x A \supset \neg \forall x \neg A$$
$$\forall x A \,\&\, \forall x B \supset \forall x(A \,\&\, B)$$
$$\exists x A \vee \exists x B \supset \exists x(A \vee B)$$

If for all x A, then there is no x such that $\neg A, \ldots,$ *If there is an x such that A or an x such that B, then there is an x such that* $A \vee B$. These sound correct. What, then, about:

$$\forall x \exists y A \supset \exists y \forall x A$$
$$\exists y \forall x A \supset \forall x \exists y A$$

It turns out that the former is not correct, but the latter is. In general, if something is not logically correct, there should be a **counterexample**. The Norwegian logician Thoralf Skolem showed in 1922, in a talk he gave in Helsingfors, that if there is a counterexample for a formula in predicate logic, there is one that talks about the natural numbers and their properties. We take as the A of the first of the above formulas the two-place atomic formula $y > x$. Now the whole formula states:

If for every number x there is a number y greater than x, then there is a number y such that for every number x, y is greater than x.

The antecedent of this implication is a clearly true statement about natural numbers, the succedent instead, claiming the existence of a greatest natural number, a clearly false statement. Therefore the first of the above implications cannot be a logically correct one.

The quantifiers in the antecedent $\forall x \exists y\ y > x$ can be taken in the following way: Whatever value is proposed for x, a value can be found for y such that $y > x$. This is easy: just set $y = x + 1$. Then $y > x$ is $x + 1 > x$ which is true whatever the value of x. With $\exists y \forall x\ y > x$, one would first have to fix a value for y, then show that in whatever way x is chosen, a formula true about the natural numbers is obtained. However, with y fixed, if x is chosen to be $y + 1$, the result is $y > y + 1$ which is a falsity whatever y was chosen.

The pattern behind the above readings of the quantifiers is as follows: Let $Q_1, Q_2, \ldots$ stand for $\forall$ or $\exists$:

A sequence of quantifiers $Q_1 x, \ldots, Q_n z$ *before a formula A codes an order of dependence between the variables x, ..., z in A such that each occurrence of an existentially quantified variable depends on the variables that stand before it in the quantifier prefix* $Q_1 x, \ldots, Q_n z$.

This idea will suffice for seeing the logical correctness of:

$$\exists y \forall x A \supset \forall x \exists y A$$

The antecedent $\exists y \forall x A$ requires that y be chosen first, and that any later choice of x will result in a correct claim A about some relation between the chosen y and x. The consequent $\forall x \exists y A$ requires that whatever x was chosen first, a y can be found such that A is correct. To meet this requirement, it is sufficient to make each time the same choice of y as in the antecedent.

As we have seen, it is by no means immediately evident whether something is a logically correct claim or logically correct argument when quantifiers are used. Similarly to propositional logic, a few logical principles of proof in predicate logic turn out to be sufficient for the representation of all logically correct arguments. This is by no means an obvious property. It turns out further that if a domain is finite, the quantifiers do not add anything essential to propositional logic. If a domain is infinite, what guarantee is there, *a priori*, that a finite system of rules is sufficient for the representation of all logically correct arguments about the domain?

8.2 The meaning of the quantifiers

There are two main theories about the meaning of universal propositions. To explain them, we need a precise notion of **instances** of such propositions. This is easy in specific cases, as with the natural numbers and a universal proposition such as *All even numbers are sums of two prime numbers.* The instances are *2 is a sum of two prime numbers, 4 is a sum of two prime numbers, ...* Formally, we do the following. First, constants and variables together are called **terms** and denoted by $s, t, \ldots$. We use a special bracket notation for the **substitution** of a term in a formula:

Substitution. *Let A be a formula. Then $A[t/x]$ is the formula that is obtained from A when the free occurrences of x in A are replaced by the term t.*

If A has no free occurrences of x, then $A[t/x]$ is identical to A. Otherwise, in the 'normal' case, it is some other formula. We say that t is **free for** x in A if no variable in t becomes bound in the substitution.

Given a universal formula $\forall x A$, an **instance** of $\forall x A$ is obtained through the substitution of a term t for x in A, i.e., $A[t/x]$ is an instance of $\forall x A$.

The first of the two theories about the meaning of universal propositions, associated with the name of Alfred Tarski in the 1930s, goes as follows:

Truth of universals: Tarski. *Given a domain $\mathcal{D}$ of objects $a_1, a_2, \ldots$ and relations $P, Q, R, \ldots$ in $\mathcal{D}$, the formula $\forall x A$ is* **true in** *$\mathcal{D}$ if the instance $A[a_i/x]$ is true for each a_i. $\forall x A$ is* **logically true** *if it is true in any domain $\mathcal{D}$.*

The idea is that, given a schematic formula A of predicate logic, it can be 'interpreted' as a concrete formula about some concrete objects in a given domain and with some concrete relations in place of the schematic atomic formulas. Thus, one says that formula A is logically true if it is true under any interpretation.

Tarski's definition is problematic in a couple of respects. First of all, it refers to an arbitrary domain of objects and some theory would be needed for stating what that means. The most common account relies on axiomatic set theory. Secondly, the definition mainly replaces the *for all* in $\forall x A$ by *each* a_i in $A[a_i/x]$. From one point of view, the definition is obtained through ellipsis: If a domain $\mathcal{D}$ is **finite**, we can list its elements, as in $a_1, \ldots, a_n$. The different instances of $\forall x A$ are $A[a_1/x], \ldots, A[a_n/x]$. Therefore $\forall x A$ is true in a finite domain $\mathcal{D}$ if each of the instances $A[a_1/x], \ldots, A[a_n/x]$ is true in $\mathcal{D}$. These instances are true in $\mathcal{D}$ if and only if their conjunction is true in $\mathcal{D}$, so we have:

A universal formula $\forall x A$ is true in a finite domain $\mathcal{D}$ of n objects $a_1, \ldots, a_n$ if $A[a_1/x]$ & $\ldots$ & $A[a_n/x]$ is true in $\mathcal{D}$.

Tarski's meaning explanation of the universal quantifier follows if the restriction to finiteness is lifted, as in:

$\forall x A$ is true in $\mathcal{D}$ if $A[a_1/x]$ & $A[a_2/x]$ & $\ldots$ is true in $\mathcal{D}$.

Another, much earlier account of the meaning of universal propositions stems from Frege in 1879. Gentzen formulated it in natural deduction as:

Provability of universals: Frege–Gentzen. *The formula $\forall x A$ is* **provable** *from assumptions Γ if $A[y/x]$ is provable for an* **arbitrary** *y.*

The Frege–Gentzen account is syntactic and does not need any theory of what domains of objects are in general. It is just assumed that there is an unbounded supply of parameters $a, b, c, \ldots$ that represent constants, as well as variables $x, y, z, \ldots$. As to 'arbitrary' variables, there is nothing arbitrary about them. We shall make the notion precise through an introduction rule for $\forall$ in natural deduction:

Table 8.4 The introduction rule for the universal quantifier

$$\frac{\begin{matrix}\Gamma \\ \vdots \\ A[y/x]\end{matrix}}{\forall x A}\,\forall I$$

We have in the premiss a derivation of the formula $A[y/x]$ from the open assumptions Γ, and the conclusion is $\forall x A$. The following condition states in what sense y is arbitrary:

Variable condition in rule $\forall I$. *The variable y must not occur free in any of the assumptions Γ the premiss $A[y/x]$ of the rule depends on.*

The idea is that nothing is assumed about y except that it is an object in a domain $\mathcal{D}$. Thus, any object in $\mathcal{D}$ can take its place in the derivation of $A[y/x]$ from Γ, say the object a. If every free occurrence of y in the derivation is replaced by a, a derivation of $A[a/x]$ from Γ is obtained. The arbitrary variable in rule $\forall I$ is called the **eigenvariable** of $\forall I$. In practice, when an inference with an eigenvariable is planned, a variable is chosen as eigenvariable that does not occur anywhere else. Such a variable is called a **fresh** variable.

The condition of provability according to Frege–Gentzen is stronger than the condition of truth according to Tarski: If we have a derivation of $A[y/x]$ from Γ for an arbitrary y, we can substitute in turn $a_1, a_2, \dots$ for x in the derivation, to get a derivation of $A[a_1/x]$ from Γ, then $A[a_2/x]$ from Γ, etc. This is not an **infinitistic** explanation like the one of Tarski, because we have a finitary prescription for how to produce a derivation of $A[a_i/x]$ from Γ for any given object a_i. The prescription is not unlike, say, the prescription for doing the sum of two natural numbers that we do not need to explain by actually showing the infinity of possible sums of two numbers.

Let the domain consist of the natural numbers $0, 1, 2, \dots$, denoted $\mathcal{N}$. Let A be a formula of arithmetic with just x as a free variable. Each instance $A[n/x]$ of $\forall x A$ states something specific about some number n. Let there be reasons **specific to** n for why $A[n/x]$ is true, and let these reasons be such that they would not apply to any other number m. For all we know about arithmetic truth, there is nothing to exclude the possibility of a true universal formula $\forall x A$ such that each instance is true for its own reasons, so to say. If this is so, there is no finitary way of seing that $\forall x A$ is true. The Frege–Gentzen provability condition instead requires there to be a justification for A that is **uniform**, the same for each instance. We can ask again, as in the last paragaph of Section 7.3: Which comes first, truth or provability?

The meaning of the existential quantifier in a sentence such as $\exists x A$ is explained in terms of truth by requiring that some instance $A[a_i/x]$ be true:

Truth for existence. *Given a domain $\mathcal{D}$ of objects $a_1, a_2, \ldots$, an existential formula $\exists x A$ is* **true in** *$\mathcal{D}$ if $A[a_i/x]$ is true for some a_i. $\exists x A$ is* **logically true** *if it is true in any domain $\mathcal{D}$.*

If a domain is finite, truth reduces to the propositional case similarly to the universal quantifier:

An existential formula $\exists x A$ is true in a finite domain $\mathcal{D}$ of n objects $a_1, \ldots, a_n$ if $A[a_1/x] \lor \ldots \lor A[a_n/x]$ is true in $\mathcal{D}$.

There is a limiting case in which the number of objects is 0 and the domain is **empty**. In this case we set:

Definition 8.2. Truth in an empty domain. *Existential formulas are false in an empty domain, universal formulas are true in an empty domain.*

The thing is clear for an existential formula. With universal formulas, one can think that they are true in an empty domain because there exist no counterexamples.

The explanation of the existential quantifier in terms of provability is that $\exists x A$ can be concluded if $A[t/x]$ had been concluded for some term t:

Table 8.5 The introduction rule for the existential quantifier

$$\frac{\begin{array}{c}\Gamma \\ \vdots \\ A[t/x]\end{array}}{\exists x A}\ \exists I$$

The reason for requiring a term, and not some constant as in the truth condition for $\exists x A$, has to do with the way free and bound variables are treated in a logical calculus. It has to be done so that what variable is chosen as a bound variable makes no difference, i.e., so that the formulas $\forall x A$ and $\exists x A$ can be changed into $\forall y A[y/x]$ and $\exists y A[y/x]$, respectively, whenever y is free for x in A. The change of bound variables goes under the name of **α-conversion**. The way to include α-conversion in a logical calculus, instead of just saying that bound variables can be changed, is to use terms in a suitable way in rule $\exists I$ and in the elimination rule for the universal quantifier.

The standard elimination rule for the universal quantifier is to draw arbitrary instances of a universally quantified formula:

Table 8.6 The elimination rule for the universal quantifier

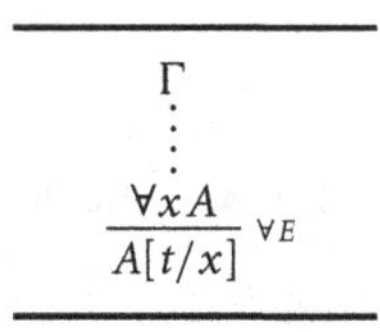

If an introduction is followed by an elimination, the *I-E* pair can be deleted, as with the following part of a derivation and its conversion:

Table 8.7 A detour conversion on $\forall$

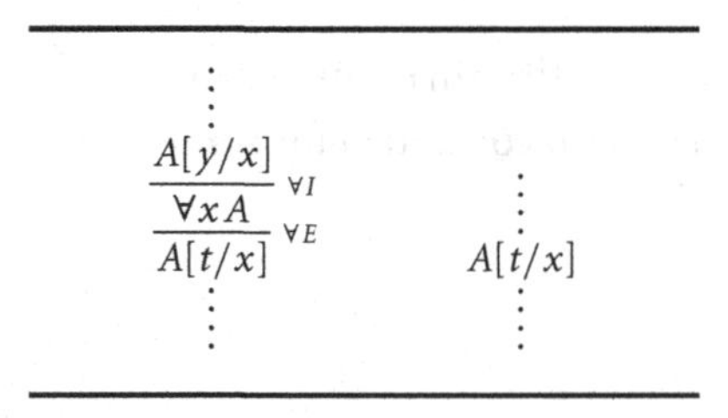

In the converted derivation, the steps down to $A[t/x]$ are obtained from the derivation of the premiss $A[y/x]$ in the unconverted derivation, through the substitution of t for y for every free occurrence of y in the latter derivation. This can be done and a correct derivation of $A[t/x]$ obtained because y is arbitrary.

The following derivation shows how the change of a bound variable in a universally quantified formula can be effected, whenever y is a fresh variable:

Table 8.8 Change of a universally bound variable

$$\dfrac{\dfrac{\forall x A}{A[y/x]}\,\forall E}{\forall y A[y/x]}\,\forall I$$

The condition on y guarantees that it works as an eigenvariable.

The elimination rule for the existential quantifier requires some preparations. In the introduction rule, any instance $A[t/x]$ can occur as a premiss, so in particular any instance $A[a/x]$ with a a constant in some domain $\mathcal{D}$. There is an analogy to disjunction: $A \vee B$ can be introduced whenever one of A and B has been derived. In the elimination rule for disjunction, a consequence C can be drawn if it follows from both of the cases that correspond to the two ways in which the disjunction can have been introduced. With existence, the cases are as many as there are members in the domain. To show that C is a consequence of each of these cases, we can use eigenvariables in a way similar to rule $\forall I$. Thus, if C follows from $A[y/x]$ in which y is an eigenvariable, it follows from $\exists x A$:

Table 8.9 The elimination rule for the existential quantifier

$$\dfrac{\begin{matrix}\Gamma \\ \vdots \\ \exists x A\end{matrix} \qquad \begin{matrix}\overset{1}{A[y/x]}, \Delta \\ \vdots \\ C\end{matrix}}{C}\,\exists E,1$$

The assumption $A[y/x]$ is closed at the inference, as indicated by the label.

Variable condition in rule $\exists E$. *The variable y must not occur free in any of the assumptions Δ that remain open in the derivation of the minor premiss C, nor in C.*

Similarly to the universal quantifier, an $\exists I$-$\exists E$ pair can be converted:

Table 8.10 A detour conversion on $\exists$

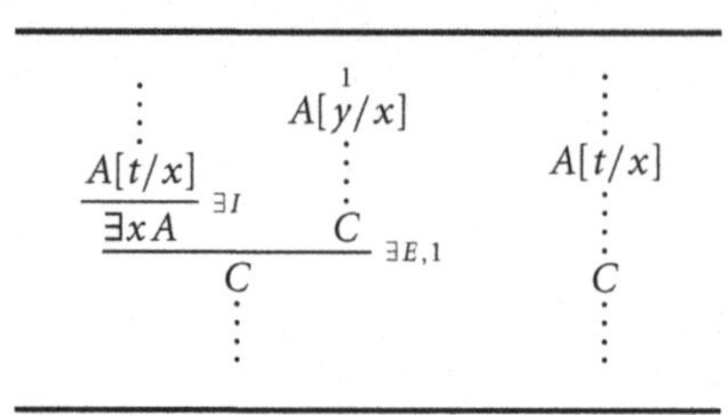

The converted derivation is obtained by taking the derivation of the premiss $A[t/x]$ of rule $\exists I$, combined with the derivation of C from $A[t/x]$ with t in place of the eigenvariable y in the latter derivation. Finally, the derivation is continued from C as in the conclusion of rule $\exists E$ in the unconverted derivation.

Variables bound by existential quantifiers can be changed similarly to the universal quantifier:

Table 8.11 Change of an existentially bound variable

$$\cfrac{\exists x A \quad \cfrac{\overset{1}{A[y/x]}}{\exists y A[y/x]}\,\exists I}{\exists y A[y/x]}\,\exists E,1$$

To show that bound variables can be changed also inside compound formulas is a matter that is best treated in a more advanced book.

The most typical error committed with the quantifier rules is: Assume $A[y/x]$. The assumptions Γ that $A[y/x]$ depends on are absent so that y cannot occur free in them, so we can generalize by $\forall I$ into $\forall x A$. The simple error lies in forgetting that the assumption $A[y/x]$ depends on itself, thus, y occurs free in an assumption $A[y/x]$ depends on. Similarly, the eigenvariable in rule $\exists E$ must not occur free in the conclusion. For example, in the change of a bound existential variable, y has to be bound before the conclusion is drawn. Without this condition, we could derive:

$$\cfrac{\exists x A \quad \overset{1}{A[y/x]}}{A[y/x]}\,\exists E,1$$

Now $\forall x A$ could be inferred from the conclusion, because y is not free in any assumption the conclusion depends on.

The addition of the above introduction and elimination rules to the rules of natural deduction of Section 3.2 gives Gentzen's original system of natural deduction for intuitionistic predicate logic. His main theorem states that derivations in the system can be converted to normal form. There are the two detour conversions with the pairs $\forall I$-$\forall E$ and $\exists I$-$\exists E$. There is in addition a permutative conversion with rule $\exists E$, in analogy to the permutative conversion for disjunction in Section 3.3, whenever the conclusion of rule $\exists E$ is the major premiss of another E-rule. For example, if also the latter rule is $\exists E$, we have a part of derivation and its conversion:

Table 8.12 A permutative conversion for rule $\exists E$

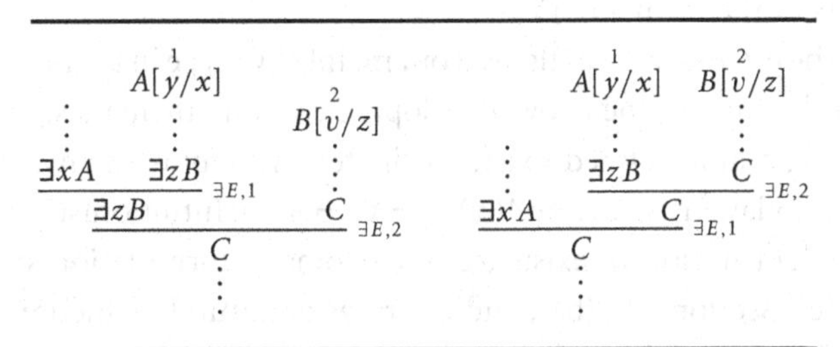

If in the unconverted derivation the minor premiss $\exists z B$ of the upper rule was derived by rule $\exists I$, the derivation after the permutative conversion has a detour convertibility on $\exists z B$.

By the normalizability of derivations, the subformula property of normal derivations follows, however, with an important change as compared to propositional logic; We have to admit as subformulas of quantified formulas any of their instances:

Table 8.13 Subformulas of a formula in predicate logic

1. *The* **subformulas** *of* $P, Q, R, \ldots, A\&B, A \vee B, A \supset B$ *without free variables, and of* $\bot$, *are these formulas themselves as well as the subformulas of* A *and* B.
2. *The* **subformulas** *of a formula* A *with a free variable* x *are* A *itself as well as the subformulas of* $A[t/x]$ *for any term* t.
3. *The* **subformulas** *of* $\forall x A$ *and* $\exists x A$ *are these formulas themselves as well as the subformulas of* A.

With an infinite domain, a quantified formula has an infinity of subformulas. Proof search cannot be limited in the way of propositional logic and the decidability of the derivability relation is lost.

Notes to Chapter 8

Intuitive reasoning with the quantifiers is at least as old as mathematics. In the latter part of the nineteenth century, there was a movement towards more rigorous formulations of mathematical arguments, especially arguments about the convergence of infinite series of rational numbers. The traditional form is that such a sequence $a_1, a_2, \ldots$ converges to a limit a if for all n, an m can be found such that for all k, if $k > m$, then $|a_k - a| < 1/n$. Intuitively, whatever small bound $1/n$ is set, there is an element a_m in the series such that all of the successive elements a_k with $k > m$ are within the bound $1/n$ from the limit value a. Thinking of it, we have here a formula with a quantifier prefix of the form $\forall x \exists y \forall z$.

The logic of the quantifiers was figured out mainly by Frege in his *Begriffsschrift*, around the time of the above developments in mathematics. Frege used classical logic and defined existence in terms of universality, in a definition that is today written as $\exists x A(x) \equiv \neg \forall x \neg A(x)$. Intuitionistic logic has an independent notion of existence. Its axiomatic formulation stems from Heyting (cf. Section 14.4(b)), and the rules of natural deduction for the two quantifiers from Gentzen.

9 | Derivations in predicate logic

We shall present first the introduction and elimination rules for the quantifiers in detail, with examples of derivations in intuitionistic predicate logic. Next a sequent system for proof search is introduced, in analogy to propositional logic. Finally, natural deduction and sequent calculus for classical predicate logic is presented.

9.1 Natural deduction for predicate logic

Some example derivations will show how the system of natural deduction for intuitionistic predicate logic works. Let us make the notation for substitution lighter by the following:

Convention about substitution. *The free variables of a formula A can be written as if the formula were atomic.*

With this convention, we write $A(x)$ for a formula A that has a free variable x, and $A(t)$ for the result $A[t/x]$ of substituting t for x in A. With more than one free variable, we write $B(x, y)$, etc. The rules of inference for the quantifiers now read:

Table 9.1 Rules of natural deduction for the quantifiers

$$\frac{A(y)}{\forall x A(x)}\,{\scriptstyle \forall I} \qquad \frac{\forall x A(x)}{A(t)}\,{\scriptstyle \forall E} \qquad \frac{A(t)}{\exists x A(x)}\,{\scriptstyle \exists I} \qquad \frac{\exists x A(x) \quad \begin{matrix} \overset{1}{A(y)} \\ \vdots \\ C \end{matrix}}{C}\,{\scriptstyle \exists E,1}$$

Rules $\forall I$ and $\exists E$ have the eigenvariable y. A system of natural deduction for intuitionistic predicate logic is obtained through the addition of these quantifier rules to the system of intuitionistic natural deduction for propositional logic of Chapter 3, and a system of classical predicate logic through an addition to the classical propositional rules of Chapter 5.

(a) Some example derivations. We shall show as a first example of derivations in natural deduction for predicate logic that the order of two $\forall$'s can be changed, and the same for two $\exists$'s:

$$\dfrac{\dfrac{\dfrac{\dfrac{\forall x \forall y A(x, y)}{\forall y A(x, y)}\,\forall E}{A(x, y)}\,\forall E}{\forall x A(x, y)}\,\forall I}{\forall y \forall x A(x, y)}\,\forall I \qquad \dfrac{\exists x \exists y A(x, y) \quad \dfrac{\overset{2}{\exists y A(x, y)} \quad \dfrac{\dfrac{\overset{1}{A(x, y)}}{\exists x A(x, y)}\,\exists I}{\exists y \exists x A(x, y)}\,\exists I}{\exists y \exists x A(x, y)}\,\exists E{,}1}{\exists y \exists x A(x, y)}\,\exists E{,}2$$

The variable condition is met in the left derivation, because there are no free variables in the assumption. In both derivations, we have used x and y as eigenvariables. This choice respects the variable conditions for the quantifier rules.

Some of the simplest types of formulas in predicate logic belong to what is known as **monadic predicate logic**, i.e., that special case of predicate logic in which there are only one-place predicates. Such predicates can be used for representing Aristotle's theory of **syllogistic inferences.** These inferences have two premisses and one conclusion, in the style of *If every A is B and every B is C, then every A is C*, and *If some A is B and every B is C, then some A is C.* These two examples can be written in the language of predicate logic as:

$$\forall x(A(x) \supset B(x)), \forall y(B(y) \supset C(y)) \vdash \forall z(A(z) \supset C(z))$$

$$\exists x(A(x) \,\&\, B(x)), \forall y(B(y) \supset C(y)) \vdash \exists z(A(z) \,\&\, C(z))$$

In these, we have chosen distinct bound variables in the assumptions and the conclusion. By the change of bound variables, exactly the same is expressed if we write, say in the first:

$$\forall x(A(x) \supset B(x)), \forall x(B(x) \supset C(x)) \vdash \forall x(A(x) \supset C(x))$$

This syllogism is derived as in:

$$\dfrac{\dfrac{\dfrac{\dfrac{\forall x(B(x) \supset C(x))}{B(x) \supset C(x)}\,\forall E \quad \dfrac{\dfrac{\forall x(A(x) \supset B(x))}{A(x) \supset B(x)}\,\forall E \quad \overset{1}{A(x)}}{B(x)}\,\supset E}{C(x)}\,\supset E}{A(x) \supset C(x)}\,\supset I{,}1}{\forall x(A(x) \supset C(x))}\,\forall I$$

There are no free variables left in the open assumptions after the second-to-last line has been concluded, so that x works as an eigenvariable in the last rule. Nothing in the eigenvariable conditions prevents the use of the bound variable also as the eigenvariable.

The second example is derived by:

$$\dfrac{\exists x(A(x)\,\&\,B(x)) \qquad \dfrac{\dfrac{\dfrac{\overset{1}{A(z)\,\&\,B(z)}}{A(z)}\&E \qquad \dfrac{\dfrac{\forall y(B(y)\supset C(y))}{B(z)\supset C(z)}\forall E \qquad \dfrac{\overset{1}{A(z)\,\&\,B(z)}}{B(z)}\&E}{C(z)}\supset E}{A(z)\,\&\,C(z)}\&I}{\exists z(A(z)\,\&\,C(z))}\exists I}{\exists z(A(z)\,\&\,C(z))}\exists E,1$$

It is impossible to tell in advance where the major premiss $\exists x(A(x)\,\&\,B(x))$ should be written. Therefore it is best just to take the instance $A(z)\,\&\,B(z)$ with the eigenvariable, to work the way to the minor premiss of $\exists E$, and then to add the major premiss and its inference line.

Note that what looks very similar to the second syllogism is not derivable:

$$\forall x(A(x)\supset B(x)), \exists y(B(y)\,\&\,C(y)) \vdash \exists z(A(z)\,\&\,C(z))$$

The failure can be seen semantically in classical logic as follows: Assume given a domain in which $A(t)$ is not true for any object. Then $A(t)\supset B(t)$ is true for any object, so that $\forall x(A(x)\supset B(x))$ is true. Assume further that there is an object a in the domain such that $B(a)$ is true and $C(a)$ is true. Then $\exists y(B(y)\,\&\,C(y))$ is true, but $\exists z(A(z)\,\&\,C(z))$ is false because $\exists z A(z)$ is false. Whatever fails classically fails also intuitionistically. With intuitionistic predicate logic, it can be shown through failed proof search that $\exists z(A(z)\,\&\,C(z))$ is underivable from the assumptions $\forall x(A(x)\supset B(x))$ and $\exists y(B(y)\,\&\,C(y))$.

Aristotle's syllogistic contains a good number of modes of inference, 19 according to one count. It is remarkable that such modes of inference, and any other similar ones, can be justified by the four quantifier rules and the rules for the connectives.

More example derivations in predicate logic:

a. $\vdash \forall x A(x) \supset \neg\exists x \neg A(x)$

$$\dfrac{\dfrac{\dfrac{\overset{2}{\exists x\neg A(x)} \qquad \dfrac{\overset{1}{\neg A(y)} \qquad \dfrac{\overset{3}{\forall x A(x)}}{A(y)}\forall E}{\bot}\supset E}{\bot}\exists E,1}{\neg\exists x\neg A(x)}\supset I,2}{\forall x A(x)\supset\neg\exists x\neg A(x)}\supset I,3$$

b. $\vdash \exists x A(x) \supset \neg\forall x \neg A(x)$

$$
\dfrac{\dfrac{\dfrac{\overset{3}{\exists x A(x)} \qquad \dfrac{\dfrac{\overset{2}{\forall x \neg A(x)}}{\neg A(y)}\,\forall E \qquad \overset{1}{A(y)}}{\bot}\,\supset E}{\bot}\,\exists E,1}{\neg\forall x \neg A(x)}\,\supset I,2}{\exists x A(x) \supset \neg\forall x \neg A(x)}\,\supset I,3
$$

The converses to (a) and (b) are underivable in intuitionistic predicate logic, but they are instead derivable in classical predicate logic. The standard interpretation of the classically provable equivalence $\exists x A(x) \supset\subset \neg\forall x \neg A(x)$ is that there is no independent notion of existence in classical logic.

c. $\vdash \forall x A(x) \,\&\, \forall x B(x) \supset \forall x(A(x) \,\&\, B(x))$

$$
\dfrac{\dfrac{\dfrac{\dfrac{\dfrac{\overset{1}{\forall x A(x) \,\&\, \forall x B(x)}}{\forall x A(x)}\,\&E}{A(y)}\,\forall E \qquad \dfrac{\dfrac{\overset{1}{\forall x A(x) \,\&\, \forall x B(x)}}{\forall x B(x)}\,\&E}{B(y)}\,\forall E}{A(y) \,\&\, B(y)}\,\&I}{\forall x(A(x) \,\&\, B(x))}\,\forall I}{\forall x A(x) \,\&\, \forall x B(x) \supset \forall x(A(x) \,\&\, B(x))}\,\supset I,1
$$

d. $\vdash \exists x(A(x) \,\&\, B(x)) \supset \exists x A(x) \,\&\, \exists x B(x)$

$$
\dfrac{\dfrac{\overset{2}{\exists x(A(x) \,\&\, B(x))} \qquad \dfrac{\dfrac{\dfrac{\overset{1}{A(y) \,\&\, B(y)}}{A(y)}\,\&E}{\exists x A(x)}\,\exists I \qquad \dfrac{\dfrac{\overset{1}{A(y) \,\&\, B(y)}}{B(y)}\,\&E}{\exists x B(x)}\,\exists I}{\exists x A(x) \,\&\, \exists x B(x)}\,\&I}{\exists x A(x) \,\&\, \exists x B(x)}\,\exists E,1}{\exists x(A(x) \,\&\, B(x)) \supset \exists x A(x) \,\&\, \exists x B(x)}\,\supset I,2
$$

In the semantic argument above, it was concluded that $\exists z(A(z) \,\&\, C(z))$ is false because $\exists z A(z)$ is false, which follows from (d).

Looking at the four example derivations, we notice that they are all normal in the simple sense of not having any detour or permutation convertibilities. However, they all have formulas such as $A(y)$ that are not strictly speaking subformulas of the conclusion (there are no open assumptions in any of the derivations). The definition of a subformula is adjusted so that any instances of universal and existential formulas are to be counted as subformulas.

It follows from the normalization theorem for intuitionistic predicate logic, to be proved in Section 13.2, that normal derivations have the subformula property. The result gives us an important corollary about the relations between the two logical systems of propositional and predicate logic, called **conservativity:**

Conservativity of predicate logic over propositional logic. *If a formula A is derivable from Γ in intuitionistic predicate logic and if there are no quantifiers in A, Γ, then A is derivable in intuitionistic propositional logic.*

Conservativity in general is the requirement that if a language is extended and new principles of proof about the extension added, there should be nothing provable about the unextended language that was not provable with the old principles.

(b) Axiomatic predicate logic. Axiomatic propositional logic was introduced in Section 3.6. Two axioms and two rules of inference are added to the propositional axioms in axiomatic predicate logic:

Table 9.2 Axiomatic predicate logic

$$\frac{C \supset A(y)}{C \supset \forall x A(x)} \qquad \forall x A(x) \supset A(t)$$

$$A(t) \supset \exists x A(x) \qquad \frac{A(y) \supset C}{\exists x A(x) \supset C}$$

The condition in the two rules is that the eigenvariable y must not occur free in C. The rules are quite analogous to those of sequent calculus, save that there is an implication in place of the turnstile.

9.2 Proof search

The examples of the previous section show again that derivation trees can be awkward to construct in natural deduction. We shall follow a path to a sequent calculus formulation of the logical rules that supports root-first proof search, similar to the one in Section 3.4: The elimination rule for $\forall$ is changed so that it has an arbitrary consequence and normal derivability is defined as in that same section. Now the rules can be translated into rules of sequent calculus that are easy to use in the construction of derivations.

(a) A modification of universal elimination. The introduction rule for $\forall$ requires that there be a derivation of $A(y)$ for an arbitrary y. Assume there to be such a derivation at hand. The eigenvariable y in it can be substituted by any term t and a correct derivation of $A(t)$ results. To express that something is the consequence of the existence of a derivation of $A(y)$ with y arbitrary, we thus stipulate:

Whatever follows from any instance $A(t)$ of $\forall x A(x)$, must follow from $\forall x A(x)$.

The corresponding elimination rule is now straightforward:

Table 9.3 The general elimination rule for $\forall$

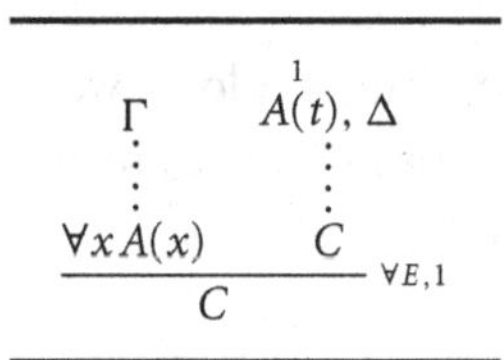

There can be any term t in the closed assumption $A(t)$. Gentzen's rule, as in Table 9.1, follows when C is identical to $A(t)$. Permutative conversions work exactly as with the existential quantifier, Table 8.10. Therefore we can define:

Definition 9.1. Normality. *A derivation in intuitionistic predicate logic with general elimination rules is* **normal** *when all major premisses of elimination rules are assumptions.*

When constructing derivations, it is often practical to use the standard elimination rules. The general rules are useful in the study of the structure of derivations. For example, it follows from normalization, as in propositional logic, that the last rule in a derivation that has no open assumptions is an I-rule. In particular, in a normal derivation of a formula $\exists x A(x)$ in which no open assumptions are left, the last rule must be $\exists I$. Therefore the following result is obtained:

Existence property. *If $\exists x A(x)$ is derivable in intuitionistic predicate logic, then $A(t)$ is derivable for some term t.*

This is a very weak result: In fact, the premiss of rule $\exists I$ is of the form $A(t)$, with no open assumptions, and nothing is assumed about t. It is just a symbol that works as well as an eigenvariable, so we could have concluded $\forall x A(x)$. To have a result with some content, let any number of atomic assumptions about objects be given, such as $P(a)$, $Q(b, c)$, etc. If the last rule in a normal derivation were an E-rule, its major premiss would be an assumption. However, it would have to be a compound formula, so that the existence property is extended:

Existence property under atomic assumptions. *If $\exists x A(x)$ is derivable from atomic assumptions Γ in intuitionistic predicate logic, then $A(t)$ is derivable from Γ for some term t.*

The premiss of the last rule, $\exists I$, can very well be $A(t)$ where t is some object known from the atoms in Γ.

The existence property can be generalized to hold under any assumptions that do not contain any essential existential quantifiers, in a way analogous to the disjunction property of intuitionistic propositional logic. For example, $\exists x A(x) \supset B$ is a formula with an existential quantifier that is only apparent, by the equivalence of $\exists x A(x) \supset B$ and $\forall x(A(x) \supset B)$. The derivation in one direction is:

$$\dfrac{\dfrac{\dfrac{\dfrac{\overset{2}{\exists x A(x) \supset B} \quad \dfrac{\overset{1}{A(x)}}{\exists x A(x)}\,{\scriptstyle \exists I}}{B}\,{\scriptstyle \supset E}}{A(x) \supset B}\,{\scriptstyle \supset I,1}}{\forall x(A(x) \supset B)}\,{\scriptstyle \forall I}}{(\exists x A(x) \supset B) \supset \forall x(A(x) \supset B)}\,{\scriptstyle \supset I,2}$$

The other direction is given by:

$$\dfrac{\dfrac{\dfrac{\overset{2}{\exists x A(x)} \quad \dfrac{\dfrac{\overset{3}{\forall x(A(x) \supset B)}}{A(y) \supset B}\,{\scriptstyle \forall E} \quad \overset{1}{A(y)}}{B}\,{\scriptstyle \supset E}}{B}\,{\scriptstyle \exists E,1}}{\exists x A(x) \supset B}\,{\scriptstyle \supset I,2}}{\forall x(A(x) \supset B) \supset (\exists x A(x) \supset B)}\,{\scriptstyle \supset I,3}$$

It is assumed that x is not free in B. If it were, x could not work as an eigenvariable in rule $\forall I$ in the first derivation, because there is at that stage still the assumption $\exists x A(x) \supset B$ left open. Similarly, y could not work as an eigenvariable in the second derivation.

The above result is a perfect analogy to the theorem of propositional logic by which we have the equivalence $(A \vee B \supset C) \supset\subset (A \supset C) \& (B \supset C)$.

The equivalence of $\exists x A(x) \supset B$ and $\forall x(A(x) \supset B)$ formalizes a mode of inference that has been used on intuitive grounds in mathematical proofs from antiquity on. A theorem can be of the form:

For all x, y, and z, if $A(x, y, z)$, then B.

It can be written in predicate logic as $\forall x \forall y \forall z(A(x, y, z) \supset B)$. The proof typically proceeds through an assumption: *Let there be given objects a, b, and c such that* $A(a, b, c)$. Here it is clearly assumed that there exist some three objects a, b, and c that satisfy some requirement expressed by $A(a, b, c)$, an assumption that is formalized in predicate logic as $\exists x \exists y \exists z A(x, y, z)$. Next B is proved from this assumption, by which $\exists x \exists y \exists z A(x, y, z) \supset B$ can be concluded. By the above predicate-logical equivalence, a proof of $\forall x \forall y \forall z(A(x, y, z) \supset B)$ is obtained.

The **given objects** $a, \ldots, c$ **such that** $A(a, \ldots, c)$ *in a theorem are the eigenvariables in the elimination of an existential assumption* $\exists x \ldots \exists z A(x, \ldots, z)$.

Another way to look at the example is: The task is to prove a theorem of the form $\forall x \forall y \forall z(A(x, y, z) \supset B)$, so assume $A(x, y, z)$ for arbitrary x, y, z. Next prove B, by which $A(x, y, z) \supset B$ can be concluded. This conclusion can be generalized into $\forall x \forall y \forall z(A(x, y, z) \supset B)$ because x, y, z were arbitrary. The formulation is slightly different because $A(x, y, z)$ is assumed for arbitrary x, y, z, whereas in the first formulation the assumption was existential.

(b) Sequent calculus rules for the quantifiers. Assume next a normal derivation in intuitionistic natural deduction to be given. We translate it into a sequent calculus derivation exactly as in Chapter 4. This is straightforward for the introduction rules: They turn into corresponding rules for introducing a formula, $\forall x A(x)$ or $\exists x A(x)$, when sequents with a formula $A(y)$ or $A(t)$ in the succedents have been derived. The left rule for $\exists$ is a direct translation of the corresponding natural elimination rule: The premiss has a derivation of C from $A(y)$ and some further assumptions, the conclusion a derivation of C from $\exists x A(x)$ and these same assumptions. The left rule for $\forall$ is similarly a direct translation of the general elimination rule of Table 9.3. Altogether, the rules to be added to the intuitionistic propositional ones of Chapter 4, Table 4.1, are four:

Table 9.4 Quantifier rules for root-first proof search

$$\frac{A(t), \forall x A(x), \Gamma \vdash C}{\forall x A(x), \Gamma \vdash C}\, L\forall \qquad \frac{\Gamma \vdash A(y)}{\Gamma \vdash \forall x A(x)}\, R\forall$$

$$\frac{A(y), \Gamma \vdash C}{\exists x A(x), \Gamma \vdash C}\, L\exists \qquad \frac{\Gamma \vdash A(t)}{\Gamma \vdash \exists x A(x)}\, R\exists$$

In rule $L\forall$, the principal formula $\forall x A(x)$ is repeated in the premiss. The reason can be seen as follows: In a natural derivation, it is possible to use a premiss $\forall x A(x)$ of rule $\forall E$ several times, each time with an instance $A(t)$, $A(s), \dots$ that is appropriate for the continuation of the derivation, as in:

$$\dfrac{\dfrac{\dfrac{\overset{1}{\forall x A(x)}}{A(a)}\,\forall E \qquad \dfrac{\overset{1}{\forall x A(x)}}{A(b)}\,\forall E}{A(a)\ \&\ A(b)}\,\&I}{\forall x A(x) \supset A(a)\ \&\ A(b)}\,\supset I,1$$

In a root-first proof search, once a particular instance has been taken, there would not be any possibility of further instances if the universal formula had been left behind somewhere down in the derivation. Consider the sequent $\forall x A(x) \vdash A(a)\ \&\ A(b)$. If rule $L\forall$ is instantiated by a without repetition, the result is $A(a) \vdash A(a)\ \&\ A(b)$ that is underivable. With $\forall x A(x)$ repeated in the antecedent of rule $L\forall$, a second instance of $L\forall$ can be taken that gives $A(a), A(b), \forall x A(x) \vdash A(a)\ \&\ A(b)$, and now rule $R\&$ gives two initial sequents:

$$\dfrac{\dfrac{\dfrac{A(a), A(b), \forall x A(x) \vdash A(a) \qquad A(a), A(b), \forall x A(x) \vdash A(b)}{A(a), A(b), \forall x A(x) \vdash A(a)\ \&\ A(b)}\,R\&}{A(a), \forall x A(x) \vdash A(a)\ \&\ A(b)}\,L\forall}{\forall x A(x) \vdash A(a)\ \&\ A(b)}\,L\forall$$

Rules $R\forall$ and $L\exists$ correspond to the two natural quantifier rules with eigenvariables. In the sequent formulation, the eigenvariable condition can be expressed as:

Eigenvariable condition. *The eigenvariables of rules $R\forall$, $L\exists$ must not occur free in the conclusions of the rules.*

Some example derivations will show how root-first proof search in predicate logic works.

Examples of root-first derivations in predicate logic:

a. $\vdash \forall x A(x) \supset \neg\exists x \neg A(x)$

$$\dfrac{\dfrac{\dfrac{\dfrac{\dfrac{A(y), \forall x A(x), \neg A(y) \vdash A(y)}{\forall x A(x), \neg A(y) \vdash A(y)}\,L\forall \qquad \bot, \forall x A(x) \vdash \bot}{\forall x A(x), \neg A(y) \vdash \bot}\,L\supset}{\forall x A(x), \exists x \neg A(x) \vdash \bot}\,L\exists}{\forall x A(x) \vdash \neg\exists x \neg A(x)}\,R\supset}{\vdash \forall x A(x) \supset \neg\exists x \neg A(x)}\,R\supset$$

b. $\forall x(A(x) \supset B(x)), \forall y(B(y) \supset C(y)) \vdash \forall z(A(z) \supset C(z))$. This is the first of the Aristotelian syllogisms. The derivation grows too broad to be printed conveniently, so some abbreviations are needed. Whenever a list of assumptions in a premiss starts to repeat formulas in the conclusion, we use dots ... to indicate them. Also, the repetition of the principal formula in rule $\forall L$ is not needed so we leave it unwritten:

$$\dfrac{\dfrac{\dfrac{\dfrac{A(z), \ldots \vdash A(z) \quad \dfrac{\dfrac{B(z), B(z) \supset C(z), \ldots \vdash B(z) \quad C(z), \ldots \vdash C(z)}{B(z), B(z) \supset C(z), \ldots \vdash C(z)}{\scriptstyle L\supset}}{B(z), \forall y(B(y) \supset C(y)) \vdash C(z)}{\scriptstyle L\forall}}{A(z), A(z) \supset B(z), \forall y(B(y) \supset C(y)) \vdash C(z)}{\scriptstyle L\supset}}{A(z), \forall x(A(x) \supset B(x)), \forall y(B(y) \supset C(y)) \vdash C(z)}{\scriptstyle L\forall}}{\forall x(A(x) \supset B(x)), \forall y(B(y) \supset C(y)) \vdash A(z) \supset C(z)}{\scriptstyle R\supset}}{\forall x(A(x) \supset B(x)), \forall y(B(y) \supset C(y)) \vdash \forall z(A(z) \supset C(z))}{\scriptstyle R\forall}$$

c. $\exists x(A(x) \,\&\, B(x)), \forall y(B(y) \supset C(y)) \vdash \exists z(A(z) \,\&\, C(z))$. This second syllogism is derived as follows:

$$\dfrac{\dfrac{\dfrac{\dfrac{A(v), B(v), \ldots \vdash B(v) \quad \dfrac{\dfrac{C(v), A(v), \ldots \vdash A(v) \quad C(v), A(v), \ldots \vdash C(v)}{C(v), A(v), B(v), \ldots \vdash A(v) \,\&\, C(v)}{\scriptstyle R\&}}{C(v), A(v), B(v), \ldots \vdash \exists z(A(z) \,\&\, C(z))}{\scriptstyle R\exists}}{A(v), B(v), B(v) \supset C(v), \forall y(B(y) \supset C(y)) \vdash \exists z(A(z) \,\&\, C(z))}{\scriptstyle L\supset}}{A(v), B(v), \forall y(B(y) \supset C(y)) \vdash \exists z(A(z) \,\&\, C(z))}{\scriptstyle L\forall}}{\dfrac{A(v) \,\&\, B(v), \forall y(B(y) \supset C(y)) \vdash \exists z(A(z) \,\&\, C(z))}{\exists x(A(x) \,\&\, B(x)), \forall y(B(y) \supset C(y)) \vdash \exists z(A(z) \,\&\, C(z))}{\scriptstyle L\exists}}{\scriptstyle L\&}$$

Easy to construct, but hard to read at times, that is a price.

It can be expected that the quantifier rules behave in ways analogous to those for conjunction and disjunction in propositional logic: Rules $R\forall$ and $\exists E$, those with the eigenvariables, are invertible. Rule $R\exists$ is not invertible and can be the source of failed proof search. Rule $L\forall$ is invertible for the trivial reason that the principal formula is repeated in the premiss:

Lemma 9.2. Inversion lemma for the quantifier rules.

(i) *If* $\Gamma \vdash \forall x A(x)$ *is derivable and* v *a fresh variable, then* $\Gamma \vdash A(v)$ *is derivable.*

(ii) *If* $\exists x A(x), \Gamma \vdash C$ *is derivable and* v *a fresh variable, then* $A(v), \Gamma \vdash C$ *is derivable.*

Proof. The proof is similar to the inversion lemma for the connectives in Section 3.5 and proceeds by induction on the height (greatest number of successive steps) in a derivation.

(i) Inversion of $\Gamma \vdash \forall x A(x)$.

Base case: If $\Gamma \vdash \forall x A(x)$ is an initial sequent, its form is $\forall x A(x), \Gamma' \vdash \forall x A(x)$ and $\forall x A(x), \Gamma' \vdash A(v)$ is derived by:

$$\frac{A(v), \forall x A(x), \Gamma' \vdash A(v)}{\forall x A(x), \Gamma' \vdash A(v)}\, L\forall$$

If $\bot$ is in Γ, then $\Gamma \vdash A(v)$ is derivable.

Inductive case: If $\forall x A(x)$ is principal in the last rule of the derivation of $\Gamma \vdash \forall x A(x)$, the derivable premiss is $\Gamma \vdash A(y)$ for some eigenvariable y. This can be substituted by v. Otherwise it is not principal. Apply then the inductive hypothesis to the premisses of the last rule, then the last rule.

(ii) Inversion of $\exists x A(x), \Gamma \vdash C$.

Base case: If $\exists x A(x), \Gamma \vdash C$ is an initial sequent and C or $\bot$ is in Γ, also $A(v), \Gamma \vdash C$ is an initial sequent. Otherwise $C \equiv \exists x A(x)$ and the sequent $A(v), \Gamma \vdash \exists x A(x)$ is derived by:

$$\frac{A(v), \Gamma \vdash A(v)}{A(v), \Gamma \vdash \exists x A(x)}\, R\exists$$

Inductive case: As in (i). QED.

In the case of the invertible propositional rules, the premisses were uniquely determined once the principal formula was fixed. In two quantifier rules, a fresh eigenvariable is chosen, but the actual choice of the symbol makes no difference so that also here the premisses are uniquely determined.

As mentioned, also rule $L\forall$ is invertible, but for a reason that is different, namely the repetition of the principal formula in the premiss. The invertibility follows from a more general result by which the antecedent of a derivable sequent can be **weakened** by the addition of any formula A, provided that it does not contain variables that clash with the eigenvariables of the derivation:

Theorem 9.3. Admissibility of weakening. *If $\Gamma \vdash C$ is derivable, then for any formula A that respects the variable condition, also $A, \Gamma \vdash C$ is derivable.*

Proof. The proof is by induction on the height of derivation. If $\Gamma \vdash C$ is an initial sequent, also $A, \Gamma \vdash C$ is. If $\Gamma \vdash C$ is concluded by a logical rule, its premisses with formula A added to each are derivable by the inductive hypothesis. The logical rule now gives the conclusion $A, \Gamma \vdash C$. QED.

By this result, if the sequent $\forall x A(x), \Gamma \vdash C$ is derivable, if follows that also $A(t), \forall x A(x), \Gamma \vdash C$ is derivable for any term t, so that rule $L\forall$ is invertible. Contrary to the other invertible quantifier rules, the choice of the term t is crucial in root-first proof search.

Similarly to rule $L\forall$, we can try to instantiate rule $R\exists$ in a root-first proof search by any term. If the sequent is $\Gamma \vdash \exists x P(x)$, we get $\Gamma \vdash P(t)$ as a possible premiss. The choice can result in an underivable sequent, even if the given sequent was derivable: For an example, the sequent $P(a) \vdash \exists x P(x)$ can be derived from $P(a) \vdash P(a)$ by rule $R\exists$. If any constant b distinct from a is chosen in the succedent, an underivable sequent $P(a) \vdash P(b)$ is obtained. Therefore rule $R\exists$ is not invertible.

The rule of contraction is likewise admissible:

Theorem 9.4. Admissibility of contraction. *If $A, A, \Gamma \vdash C$ is derivable, also $A, \Gamma \vdash C$ is derivable.*

Proof. The proof is a continuation of the corresponding proof for the propositional part in Section 4.5. It goes by induction on the length of the contraction formula, with a subinduction on the height of derivation of the derivable sequent with a duplication.

Contraction of $\forall x A(x), \forall x A(x)$: If $\forall x A(x), \forall x A(x), \Gamma \vdash C$ is an initial sequent, also $\forall x A(x), \Gamma \vdash C$ is. Otherwise consider the last rule in the derivation. If $\forall x A(x)$ is not principal in the last rule, apply contraction to the premisses. If $\forall x A(x)$ is principal, the premiss is given by a sequent $A(t), \forall x A(x), \forall x A(x), \Gamma \vdash C$ for some term t. Apply the rule of contraction to get $A(t), \forall x A(x), \Gamma \vdash C$, then rule $L\forall$ to conclude $\forall x A(x), \Gamma \vdash C$.

Contraction of $\exists x A(x), \exists x A(x)$: Apply the inversion lemma to conclude that if $\exists x A(x), \exists x A(x), \Gamma \vdash C$ is derivable, also $A(y), \exists x A(x), \Gamma \vdash C$ is derivable, with y a fresh variable, then the inversion lemma again to conclude that $A(y), A(z), \Gamma \vdash C$ is derivable, with z a fresh variable. Now substitute y for z to get a derivation of $A(y), A(y), \Gamma \vdash C$, then contraction on the shorter formula $A(y)$ to get $A(y), \Gamma \vdash C$. Rule $L\exists$ gives now the contracted conclusion $\exists x A(x), \Gamma \vdash C$. QED.

As with propositional logic, sequents can be considered the same if they contain the same formulas, irrespective of multiplicity:

If in root-first proof search a sequent with a duplication is found, say, the sequent $A, A, \Gamma \vdash C$, and if the sequent $A, \Gamma \vdash C$ already appeared further *down in the search, the rule that produced the duplication need not be applied, i.e., proof search from that rule on failed.*

Predicate logic, in contrast to propositional logic, does not have the property of termination of proof search. Proof search will of course terminate in many cases, with derivability or underivability as a result, but not always. The origin of nontermination lies in the two quantifier rules $L\forall$ and $R\exists$: If a sequent is derivable, the correct instances of these rules are there to be found, but it cannot be said in advance for how long one has to try. If a sequent is underivable, ever new instances can be produced and the search need not terminate. In example (a) below, an easy pattern emerges: A formula $A(t)$ is added to the antecedent of a sequent, always with a new instance, and this keeps repeating. However, there is no way of telling in advance what the pattern to look for might be, neither is there any bound on how complicated it can be.

(c) Underivability results. We can now show underivability results in intuitionistic predicate logic:

a. $\nvdash \neg\forall x \neg A(x) \supset \exists x A(x)$

$$
\begin{prooftree}
\AxiomC{$A(y), A(z), \neg\forall x \neg A(x) \vdash \bot$}
\RightLabel{$R\supset$}
\UnaryInfC{$A(y), \neg\forall x \neg A(x) \vdash \neg A(z)$}
\RightLabel{$R\forall$}
\UnaryInfC{$A(y), \neg\forall x \neg A(x) \vdash \forall x \neg A(x)$}
\AxiomC{$\bot, A(y) \vdash \bot$}
\RightLabel{$L\supset$}
\BinaryInfC{$A(y), \neg\forall x \neg A(x) \vdash \bot$}
\RightLabel{$R\supset$}
\UnaryInfC{$\neg\forall x \neg A(x) \vdash \neg A(y)$}
\RightLabel{$R\forall$}
\UnaryInfC{$\neg\forall x \neg A(x) \vdash \forall x \neg A(x)$}
\AxiomC{$\bot \vdash \exists x A(x)$}
\RightLabel{$L\supset$}
\BinaryInfC{$\neg\forall x \neg A(x) \vdash \exists x A(x)$}
\RightLabel{$R\supset$}
\UnaryInfC{$\vdash \neg\forall x \neg A(x) \supset \exists x A(x)$}
\end{prooftree}
$$

The next-to-last step could have been $R\exists$, because both of the two applicable rules are noninvertible. Had that rule been chosen, the one above it would have been $L\supset$ with no difference in the branch in which the search continues. With $R\forall$, a fresh variable y was chosen. Higher up, another fresh variable z was chosen. It is seen that the search necessarily produces sequents that are identical save for the addition of a copy of $A(v)$ with some fresh variable v. Therefore the proof search cannot terminate. The result, together with earlier results for intuitionistic propositional logic, suggests the following:

Independence of the connectives and quantifiers: *All of the connectives* $\&, \vee, \supset$, *the formula* $\bot$, *as well as the two quantifiers* $\forall$ *and* $\exists$ *are needed in intuitionistic predicate logic.*

A conclusive proof of this result is not easy.

b. $\nvdash \neg\neg\forall x(A(x) \vee \neg A(x))$

$$\dfrac{\dfrac{\dfrac{\dfrac{\dfrac{\dfrac{\dfrac{\dfrac{\dfrac{A(z),\, A(y),\, \neg\forall x(A(x) \vee \neg A(x)) \vdash \bot}{A(y),\, \neg\forall x(A(x) \vee \neg A(y) \vdash \neg A(z)}\,{\scriptstyle R\supset}}{A(y),\, \neg\forall x(A(x) \vee \neg A(x)) \vdash A(z) \vee \neg A(z)}\,{\scriptstyle R\vee}}{A(y),\, \neg\forall x(A(x) \vee \neg A(x)) \vdash \forall x(A(x) \vee \neg A(x))}\,{\scriptstyle R\forall} \qquad \bot \vdash \bot}{A(y),\, \neg\forall x(A(x) \vee \neg A(x)) \vdash \bot}\,{\scriptstyle L\supset}}{\neg\forall x(A(x) \vee \neg A(x)) \vdash \neg A(y)}\,{\scriptstyle R\supset}}{\neg\forall x(A(x) \vee \neg A(x)) \vdash A(y) \vee \neg A(y)}\,{\scriptstyle R\vee}}{\neg\forall x(A(x) \vee \neg A(x)) \vdash \forall x(A(x) \vee \neg A(x))}\,{\scriptstyle R\forall} \qquad \bot \vdash \bot}{\neg\forall x(A(x) \vee \neg A(x)) \vdash \bot}\,{\scriptstyle L\supset}}{\vdash \neg\neg\forall x(A(x) \vee \neg A(x))}\,{\scriptstyle R\supset}$$

There are only two places of choice here: On line four from below, one of the right disjunction rules has to be instantiated or else there is a loop. If it is the first, we have $A(y)$ in the succedent of the premiss, and have to apply $L\supset$ next, which gives a loop. On line three from below, we chose a variable y different from x, and similarly higher up z different from y. The latter was dictated by the eigenvariable condition, and it is seen that one never gets instances of A with the same eigenvariable at left and at right.

The underivability of $\neg\neg\forall x(A(x) \vee \neg A(x))$ by intuitionistic principles was first seen by L. Brouwer in the late 1920s, through a counterexample that used his theory of constructive real numbers. Well knowing that the corresponding law for propositional logic, $\neg\neg(A \vee \neg A)$, is instead derivable, one wonders how Brouwer came to see the failure of the 'infinitistic excluded middle' in the first place. A syntactic proof of underivability was found by Gentzen in 1933, but he never published it and the underivability was rediscovered in the 1950s. Its importance is that it shows the following:

Failure of Glivenko's theorem in classical predicate logic. *There are classically provable negative formulas of predicate logic that are not intuitionistically provable.*

A final example shows yet another intuitionistic limitation to classical logic:

c. $\nvdash (\forall x A(x) \supset B) \supset \exists x(A(x) \supset B)$

$$\dfrac{\dfrac{\dfrac{\forall x A(x) \supset B \vdash A(y)}{\forall x A(x) \supset B \vdash \forall x A(x)}\,{\scriptstyle R\forall} \qquad B \vdash \exists x(A(x) \supset B)}{\forall x A(x) \supset B \vdash \exists x(A(x) \supset B)}\,{\scriptstyle L\supset}}{\vdash (\forall x A(x) \supset B) \supset \exists x(A(x) \supset B)}\,{\scriptstyle R\supset}$$

Rule $L\supset$ gives a loop in the left branch. The other choice for the next-to-last rule is $\exists R$:

$$\cfrac{\cfrac{\cfrac{\cfrac{A(t), \forall x A(x) \supset B \vdash \forall x A(x) \quad B, A(t) \vdash A(y)}{A(t), \forall x A(x) \supset B \vdash A(y)}\,L\supset}{A(t), \forall x A(x) \supset B \vdash \forall x A(x)}\,R\forall \qquad B, A(t) \vdash B}{\cfrac{A(t), \forall x A(x) \supset B \vdash B}{\cfrac{\forall x A(x) \supset B \vdash A(t) \supset B}{\forall x A(x) \supset B \vdash \exists x(A(x) \supset B)}\,R\exists}\,R\supset}\,L\supset}{}$$

Note how the eigenvariable condition in rule $R\forall$ prevents to instantiate with t at right. Thus, a loop is produced in the left premiss of the uppermost rule, and even the right premiss is underivable because it has no structure.

In general, a formula of predicate logic is said to be in **prenex normal form** if it has the quantifiers at the head of the formula. We shall see in the next section that all formulas have such an equivalent in classical predicate logic, whereas the above example shows the following:

Failure of prenex normal form in intuitionistic predicate logic. *Not all formulas of intuitionistic logic have an equivalent in prenex normal form.*

9.3 Classical predicate logic

We begin with some examples that show how classical natural deduction works in the case of predicate logic. Then we show that the rule of indirect proof works with the quantifier elimination rules in the same way as with the propositional ones: Whenever the major premiss of an elimination has been concluded by rule DN, the latter can be permuted down. Lastly, we turn to a completely different idea for a classical logical calculus, namely, the multisuccedent sequent calculus that is suited for root-first proof search.

(a) Natural deduction for classical predicate logic. The system of classical natural deduction for predicate logic is obtained by adding the four quantifier rules to classical propositional logic.

We shall go through some examples of classical derivations, including two of the previous section that failed in intuitionistic logic:

a. $\vdash \forall x(A(x) \vee \neg A(x))$

$$
\begin{prooftree}
\AxiomC{$\overset{2}{\neg(A(x) \vee \neg A(x))}$}
\AxiomC{$\overset{2}{\neg(A(x) \vee \neg A(x))}$}
\AxiomC{$\overset{1}{A(x)}$}
\RightLabel{$\vee I$}
\UnaryInfC{$A(x) \vee \neg A(x)$}
\RightLabel{$\supset E$}
\BinaryInfC{$\bot$}
\RightLabel{$\supset I,1$}
\UnaryInfC{$\neg A(x)$}
\RightLabel{$\vee I$}
\UnaryInfC{$A(x) \vee \neg A(x)$}
\RightLabel{$\supset E$}
\BinaryInfC{$\bot$}
\RightLabel{$DN,2$}
\UnaryInfC{$A(x) \vee \neg A(x)$}
\RightLabel{$\forall I$}
\UnaryInfC{$\forall x(A(x) \vee \neg A(x))$}
\end{prooftree}
$$

The next example corresponds to the principle of indirect existence proofs:

b. $\neg\forall x \neg A(x) \vdash \exists x A(x)$

$$
\begin{prooftree}
\AxiomC{$\neg\forall x \neg A(x)$}
\AxiomC{$\overset{2}{\neg\exists x A(x)}$}
\AxiomC{$\overset{1}{A(x)}$}
\RightLabel{$\exists I$}
\UnaryInfC{$\exists x A(x)$}
\RightLabel{$\supset E$}
\BinaryInfC{$\bot$}
\RightLabel{$\supset I,1$}
\UnaryInfC{$\neg A(x)$}
\RightLabel{$\forall I$}
\UnaryInfC{$\forall x \neg A(x)$}
\RightLabel{$\supset E$}
\BinaryInfC{$\bot$}
\RightLabel{$DN,2$}
\UnaryInfC{$\exists x A(x)$}
\end{prooftree}
$$

The converse is derivable in intuitionistic logic, so that existence can be defined classically as a negative property: for all x, any counterexample to $A(x)$ fails.

c. $\vdash (\forall x A(x) \supset B) \supset \exists x(A(x) \supset B)$

$$
\begin{prooftree}
\AxiomC{$\overset{3}{\neg(\exists x(A(x) \supset B))}$}
\AxiomC{$\overset{4}{\forall x A(x) \supset B}$}
\AxiomC{$\overset{3}{\neg(\exists x(A(x) \supset B))}$}
\AxiomC{$\overset{2}{\neg A(x)}$}
\AxiomC{$\overset{1}{A(x)}$}
\RightLabel{$\supset E$}
\BinaryInfC{$\bot$}
\RightLabel{$\bot E$}
\UnaryInfC{B}
\RightLabel{$\supset I,1$}
\UnaryInfC{$A(x) \supset B$}
\RightLabel{$\exists I$}
\UnaryInfC{$\exists x(A(x) \supset B)$}
\RightLabel{$\supset E$}
\BinaryInfC{$\bot$}
\RightLabel{$DN,2$}
\UnaryInfC{$A(x)$}
\RightLabel{$\forall I$}
\UnaryInfC{$\forall x A(x)$}
\RightLabel{$\supset E$}
\BinaryInfC{B}
\RightLabel{$\supset I$}
\UnaryInfC{$A(x) \supset B$}
\RightLabel{$\exists I$}
\UnaryInfC{$\exists x(A(x) \supset B)$}
\RightLabel{$\supset E$}
\BinaryInfC{$\bot$}
\RightLabel{$DN,3$}
\UnaryInfC{$\exists x(A(x) \supset B)$}
\RightLabel{$\supset I,4$}
\UnaryInfC{$(\forall x A(x) \supset B) \supset \exists x(A(x) \supset B)$}
\end{prooftree}
$$

The derivation clearly shows that natural deduction is not suited to proof search. It is essential that $\exists x(A(x) \supset B)$ is concluded indirectly.

Whenever the conclusion of rule *DN* is a major premiss of an elimination, it can be permuted down. Let us treat here the case of $\forall E$ for which the derivation and its transformation are:

$$
\dfrac{\dfrac{\begin{matrix}\overset{1}{\neg\forall x A(x)} \\ \vdots \\ \bot\end{matrix}}{\forall x A(x)}{\scriptstyle DN,1}}{A(t)}{\scriptstyle \forall E}
\quad\rightsquigarrow\quad
\dfrac{\begin{matrix}\dfrac{\dfrac{\overset{2}{\neg A(t)} \quad \dfrac{\overset{1}{\forall x A(x)}}{A(t)}{\scriptstyle \forall E}}{\bot}{\scriptstyle \supset E}}{\neg\forall x A(x)}{\scriptstyle \supset I,1} \\ \vdots \\ \bot\end{matrix}}{A(t)}{\scriptstyle DN,2}
$$

It turns out, instead, that *DN* does not permute with rule $\forall I$. The derivation is:

$$
\dfrac{\dfrac{\begin{matrix}\overset{1}{\neg A(y)} \\ \vdots \\ \bot\end{matrix}}{A(y)}{\scriptstyle DN,1}}{\forall x A(x)}{\scriptstyle \forall I}
$$

Trying to permute down *DN* similarly to the previous case with $\forall E$, we assume $\neg\forall x A(x)$, and should then derive a contradiction, i.e., $\forall x A(x)$, but it doesn't succeed because the variable condition blocks the generalization from $A(y)$.

(b) Proof search in classical predicate logic. A sequent calculus for classical predicate logic results if the single-succedent intuitionistic calculus is generalized so that it has any number of cases in the succedent. The propositional rules are those of Section 6.2, with the two rules of primitive negation included. The quantifier rules are as follows:

Table 9.5 Classical quantifier rules for proof search

$$
\dfrac{A(t), \forall x A(x), \Gamma \vdash \Delta}{\forall x A(x), \Gamma \vdash \Delta}{\scriptstyle L\forall}
\qquad
\dfrac{\Gamma \vdash \Delta, A(y)}{\Gamma \vdash \Delta, \forall x A(x)}{\scriptstyle R\forall}
$$

$$
\dfrac{A(y), \Gamma \vdash \Delta}{\exists x A(x), \Gamma \vdash \Delta}{\scriptstyle L\exists}
\qquad
\dfrac{\Gamma \vdash \Delta, \exists x A(x), A(t)}{\Gamma \vdash \Delta, \exists x A(x)}{\scriptstyle R\exists}
$$

The variable conditions are as before: The eigenvariables of rules $R\forall$ and $L\exists$ must not occur free in the conclusions of these rules. The left rules are just like the rules of the previous section, save for Δ as a succedent instead of a single formula. Rules $R\forall$ and $R\exists$ have similarly an added context Δ, but the latter rule has also a repetition of the principal formula.

The classical rules show a perfect mirror-like duality of the universal and existential quantifier: Rule $R\exists$ is like rule $L\forall$ if the latter is read from right to left, and similarly for rules $L\exists$ and $R\forall$. We can conclude without further work that rule $R\exists$ is invertible because the premiss is obtainable from the conclusion by weakening in the succedent.

More generally, weakening is divided into weakening **at left** and **at right**, and similarly for contraction. There was no need to even consider contraction in the classical propositional calculus, because each premiss of a rule was simpler than the conclusion. Here, instead, rules $L\forall$ and $R\exists$ have premisses that are more complex than the conclusion. These two rules are thus the source of possible nontermination.

The classical quantifier rules $L\forall$ and $R\exists$ can be used without a repetition of the principal formula in a premiss. The reason is that such rules are admissible. To show this, assume there to be a derivation of the sequent $\Gamma \vdash \Delta, A(t)$. The conclusion of rule $R\exists$ is derived as in:

$$\cfrac{\cfrac{\Gamma \vdash \Delta, A(t)}{\Gamma \vdash \Delta, \exists A(x), A(t)}\,Wk}{\Gamma \vdash \Delta, \exists A(x)}\,R\exists$$

Weakening is an admissible rule, and therefore also rule $R\exists$ without the repetition is admissible. The same holds for rule $L\forall$, but it should be kept in mind that the rules without repetition can also lead to a failure of proof search.

Let us take some examples of classical derivations:

a. $\vdash \neg\forall x \neg A(x) \supset \exists x A(x)$

$$\cfrac{\cfrac{\cfrac{\cfrac{\cfrac{A(y) \vdash A(y)}{A(y) \vdash \exists x A(x)}\,R\exists}{\vdash \neg A(y), \exists x A(x)}\,R\neg}{\vdash \forall x \neg A(x), \exists x A(x)}\,R\forall}{\neg\forall x \neg A(x) \vdash \exists x A(x)}\,L\neg}{\vdash \neg\forall x \neg A(x) \supset \exists x A(x)}\,R\supset$$

We proved this result in classical natural deduction in the end of Section 9.1, but the root-first proof is much easier to find. The converse implication was proved already for the intuitionistic calculus, so that existence can be defined in terms of the universal quantifier in classical logic:

There is no independent notion of existence in classical predicate logic.

By the connection between implication and disjunction, the above example is equivalent to $\vdash \forall x \neg A(x) \vee \exists x A(x)$. A typical intuitive application of this principle in a mathematical proof proceeds as follows: *To prove that there is an x such that* $A(x)$, *assume* $\neg A(x)$ *for all x. This turns out impossible, so we have proved the claim.* This **indirect existence proof** is purely classical and it could equally well have been formulated as an application of the law of double negation to an existential claim, namely the implication: $\neg\neg\exists x A(x) \supset \exists x A(x)$.

We can give a reading to the double negation that shows why the above step is purely classical. Let an algorithm be given for the computation of a decimal expansion, and let x range over the natural numbers with $A(n)$ expressing: *The nth decimal is nonzero.* The double negation $\neg\neg\exists x A(x)$ states: *It is impossible that there should not be a decimal that is nonzero.* It can very well happen that the proof of this fact gives no information on how many decimal places have to be computed before a nonzero decimal is found. Thus, $\neg\neg\exists x A(x)$ can be read as: *In the computation of the decimal expansion, a nonzero decimal will eventually turn out.* The difference to the direct existential claim is that its constructive proof gives a method for actually finding in a predictable number of steps a nonzero decimal.

The quantifiers $\forall$ and $\exists$ are duals to each other, so we have also:

b. $\vdash \neg\exists x \neg A(x) \supset \forall x A(x)$

$$
\cfrac{\cfrac{\cfrac{\cfrac{\cfrac{A(y) \vdash A(y)}{\vdash \neg A(y), A(y)}\,R\neg}{\vdash \exists x \neg A(x), A(y)}\,R\exists}{\vdash \exists x \neg A(x), \forall x A(x)}\,R\forall}{\neg\exists x \neg A(x) \vdash \forall x A(x)}\,L\neg}{\vdash \neg\exists x \neg A(x) \supset \forall x A(x)}\,R\supset
$$

Note that there was no need to repeat the principal formula in rule $R\exists$. This rule has to be applied before $R\forall$ to respect the eigenvariable condition. Without the condition, one could conclude $\vdash \neg A(y), \forall x A(x)$, and then $\vdash \forall x \neg A(x), \forall x A(x)$, and finally the clear fallacy $\vdash \forall x \neg A(x) \vee \forall x A(x)$.

c. $\vdash (\forall x A(x) \supset B) \supset \exists x(A(x) \supset B)$

$$\dfrac{\dfrac{\dfrac{\dfrac{A(y) \vdash A(y),\, B}{\vdash A(y),\, A(y) \supset B}\,{\scriptstyle R\supset}}{\vdash A(y),\, \exists x(A(x) \supset B)}\,{\scriptstyle R\exists}}{\vdash \forall x A(x),\, \exists x(A(x) \supset B)}\,{\scriptstyle R\forall} \qquad \dfrac{\dfrac{B,\, A(y) \vdash B}{B \vdash A(y) \supset B}\,{\scriptstyle R\supset}}{B \vdash \exists x(A(x) \supset B)}\,{\scriptstyle R\exists}}{\dfrac{\forall x A(x) \supset B \vdash \exists x(A(x) \supset B)}{\vdash (\forall x A(x) \supset B) \supset \exists x(A(x) \supset B)}\,{\scriptstyle R\supset}}\,{\scriptstyle L\supset}$$

Repetitions of principal formulas have again been left out. The converse was proved in the intuitionistic calculus. There also the equivalence of $\exists x A(x) \supset B$ and $\forall x(A(x) \supset B)$ was proved. These two results are classical duals, as is seen by writing them one below the other:

$$\vdash (\forall x A(x) \supset B) \supset\subset \exists x(A(x) \supset B)$$
$$\vdash (\exists x A(x) \supset B) \supset\subset \forall x(A(x) \supset B)$$

In classical predicate logic, also the following equivalences are derivable:

$$\vdash (A \supset \forall x B(x)) \supset\subset \forall x(A \supset B(x))$$
$$\vdash (A \supset \exists x B(x)) \supset\subset \exists x(A \supset B(x))$$

If x happens to be free in A (resp. in B in the upper two) in the left part of the equivalences, it can be substituted by a fresh variable. Then the four equivalences can be seen as equally many ways of bringing a quantifier in the head of a formula.

Definition 9.5. Prenex normal form. *A formula is in* **prenex normal form** *if it begins with a string of quantifiers followed by a quantifier-free part.*

The transformation to prenex normal form goes through all the way, not just for the case of an implication: If a universal quantification is a conjunct, we have $\forall x A(x) \,\&\, B$ and this is easily shown equivalent to $\forall x(A(x) \,\&\, B)$. In cases such as $\exists x A(x) \,\&\, \exists x B(x)$, the equivalent prenex normal form is, with a change of a bound variable occurrence, $\exists x \exists y(A(x) \,\&\, B(y))$. With negation, the transformation into prenex normal form is one of:

$$\neg\forall x A(x) \text{ becomes } \exists x \neg A(x), \qquad \neg\exists x A(x) \text{ becomes } \forall x \neg A(x).$$

By the above transformations we have:

Theorem 9.6. Prenex normal form theorem. *Each formula in classical predicate logic can be transformed into an equivalent formula in prenex normal form.*

If a formula is in prenex normal form and if it is derivable, the quantifier rules must necessarily come last in the derivation, and only two of them, the right ones, are applied. There is no mystery here: Rule $R\exists$ does, by the duality of $\forall$ and $\exists$, the work of $L\forall$. The part with quantifier rules in the derivation is preceded by a part in which only propositional rules are applied. The conclusion of the last propositional rule, at the same time the premiss of the first quantifier rule, is the **midsequent** of the derivation.

The derivability of any sequent $\Gamma \vdash \Delta$ can be reduced by the prenex normal form to the derivability of a suitable midsequent by the rules of propositional logic: Take a derivable sequent $\Gamma \vdash \Delta$, and let $\&\Gamma$ stand for the conjunction of the formulas in Γ and $\vee\Delta$ for the disjunction of the formulas in Δ. Then $\Gamma \vdash \Delta$ is derivable if and only if $\vdash \&\Gamma \supset \vee\Delta$ is derivable. The formula $\&\Gamma \supset \vee\Delta$ can be brought into prenex normal form and the midsequent theorem applied. Contrary to the good hopes that this result might give rise to, because of the decidability of derivability by the propositional rules, it is not possible to determine in advance how complicated the midsequent has to be: No bound can be set on how many instantiations by rule $R\exists$ are required in a root-first proof search.

Let us now return to weakening and contraction in the classical calculus, briefly discussed in the beginning of this section:

Theorem 9.7. Admissibility of left and right weakening.

(i) *If* $\Gamma \vdash \Delta$ *is derivable, also* $A, \Gamma \vdash \Delta$ *is derivable for any formula* A.
(ii) *If* $\Gamma \vdash \Delta$ *is derivable, also* $\Gamma \vdash \Delta, A$ *is derivable for any formula* A.

Proof. The proof is by induction on the height of derivation and goes through analogously to the proof for the intuitionistic calculus. QED.

Theorem 9.8. Admissibility of left and right contraction.

(i) *If* $A, A, \Gamma \vdash \Delta$ *is derivable, then* $A, \Gamma \vdash \Delta$ *is derivable.*
(ii) *If* $\Gamma \vdash \Delta, A, A$ *is derivable, then* $\Gamma \vdash \Delta, A$ *is derivable.*

Proof. The propositional part is different from the intuitionistic calculus so that the proof of admissibility of (left) contraction in Section 3.5 does not apply. However, we proved in Section 6.2 that all of the propositional rules are invertible. Thus, if there is a duplication of a formula, both copies can be decomposed and contraction applied to the components. From the contracted sequents, the contracted conclusion can be reached. There are eight cases. The first is contraction of $A \& B, A \& B$ in the succedent.

Decomposition goes as follows:

$$\cfrac{\cfrac{\Gamma \vdash \Delta, A, A \quad \Gamma \vdash \Delta, B, A}{\Gamma \vdash \Delta, A \& B, A}{\scriptstyle R\&} \qquad \cfrac{\Gamma \vdash \Delta, A, B \quad \Gamma \vdash \Delta, B, B}{\Gamma \vdash \Delta, A \& B, B}{\scriptstyle R\&}}{\Gamma \vdash \Delta, A \& B, A \& B}{\scriptstyle R\&}$$

The leftmost uppermost sequent is contracted into $\Gamma \vdash \Delta, A$ and the rightmost uppermost similarly to $\Gamma \vdash \Delta, B$. Rule $R\&$ gives now the conclusion $\Gamma \vdash \Delta, A \& B$. The rest of the propositional rules are handled similarly.

For the quantifier rules, contraction at left goes through as for the intuitionistic calculus, with an arbitrary context in the succedent instead of a single formula. Contraction at right is proved dually. QED.

As in the intuitionistic calculus, whenever a step of root-first proof search produces a sequent that is identical to a previous one, save for the multiplication of some formulas, proof search with that step fails. Thus, the possible sources of nontermination come from ever new instances of existential formulas in the succedent and universal formulas in the antecedent. The overall result is:

There is no general method for deciding derivability in classical predicate logic.

This result was proved by Alonzo Church in 1936. He showed that the decidability of predicate logic would result in a complete system of arithmetic, contrary to Gödel's incompleteness result. Another proof of undecidability used what is called the **halting problem for Turing machines**. It was found by Alan Turing and published in his path-breaking article of 1936. During the Second World War, Turing put his machine into practice in his decoding work (ENIGMA), and he and Max Newman began the development that led to some of the first computers and programming languages, both direct descendants of the logical ideas of computability and formal languages, respectively.

Notes and exercises to Chapter 9

1. Give derivations of the following:
   a. $\forall x A(x) \& \forall x B(x) \supset\subset \forall x(A(x) \& B(x))$
   b. $\exists x(A(x) \vee B(x)) \supset\subset \exists x A(x) \vee \exists B(x)$
   c. $\forall x(A(x) \supset B(x)) \supset (\forall x A(x) \supset \forall x B(x))$
   d. $\forall x A(x) \vee \forall x B(x) \supset \forall x(A(x) \vee B(x))$
   e. $\exists x(A(x) \& B(x)) \supset \exists x A(x) \& \exists x B(x)$
   f. $\exists x A(x) \& \exists x B(x) \supset\subset \exists x \exists y(A(x) \& B(y))$

2. Show through counterexamples that the converses to the implications in (d) and (e) do not hold.

3. Do the permutative conversion for $\forall$.

4. Show that $\neg\exists x A(x) \supset \forall x \neg A(x)$ is intuitionistically derivable. Show that $\neg\forall x A(x) \supset \exists x \neg A(x)$ instead is not intuitionistically derivable.

5. As noted, the formula $\neg\neg\forall x(A(x) \lor \neg A(x))$ is not intuitionistically derivable. Show that $\forall x \neg\neg(A(x) \lor \neg A(x))$ instead is derivable. It follows that $\forall x \neg\neg(A(x) \lor \neg A(x)) \supset \neg\neg\forall x(A(x) \lor \neg A(x))$ is not intuitionistically derivable. Show directly that even the more general law of 'double-negation shift' $\forall x \neg\neg A(x) \supset \neg\neg\forall x A(x)$ is not intuitionistically derivable.

6. **Independence of the intuitionistic quantifiers.** A detailed proof of the independence of the intuitionistic connectives and quantifiers is given in Prawitz (1965, pp. 59–62) within natural deduction. With P an atomic formula, show that if C contains no $\forall$, $\vdash \forall x P(x) \supset\subset C$ is underivable in intuitionistic sequent calculus. Similarly, if C contains no $\exists$, $\vdash \exists x P(x) \supset\subset C$ is underivable. Warning: As with $\lor$ in exercise 5 of Chapter 4, Harrop's theorem needs to be used for the latter.

10 | The semantics of predicate logic

Similarly to classical propositional logic, the classical form of predicate logic has a simple semantics. Kripke semantics for intuitionistic predicate logic instead has the complication that the domain of individual objects is not given once and for all.

10.1 Interpretations

(a) The semantics of classical predicate logic. The semantics of classical propositional logic presented in Section 7.1 was based on the idea that in each concrete situation, the truth values of atomic formulas are determined. The formal presentation was in terms of valuations, i.e., assignments of truth values to the atomic formulas. In predicate logic we have a domain of individuals and the atomic formulas make statements about the properties of individuals and relations among them. The basic ideas of the semantics of classical predicate logic were already given in Section 8.2: As explained there, the schematic atomic formulas get interpreted in a given domain, and a universal formula $\forall x A(x)$ is true under an interpretation if each of its instances $A(a)$, $A(b)$, $A(c)$, ... is true, and an existential formula $\exists x A(x)$ is similarly true if there is some instance $A(a)$ that is true under an interpretation.

If a domain is infinite, as in the case of the natural numbers, it is not possible to go through all the instances of a formula $A(x)$ with a free variable. In the Frege–Gentzen explanation of the universal quantifier in Section 8.2, provability of $\forall x A(x)$ required a proof of $A(y)$ for an arbitrary y. This condition is stronger than the truth condition for a universal formula.

The truth of an existential formula $\exists x A(x)$ under a given interpretation requires some individual a such that $A(a)$ and is analogous to the condition of provability.

Let us consider an example, the formula $\forall x \exists y A(x, y)$ that we encountered already in Section 8.1. Let the domain consist of the natural numbers $\mathcal{N}$ and let $A(x, y)$ be the relation $y > x$. Under this interpretation, $\forall x \exists y\, y > x$ is true if $\exists y\, y > n$ is true for whatever value is given to n. The latter, in turn, is true if, whatever value is given to n, there is some m such that $m > n$ is true. If we set $m = n + 1$, we get $n + 1 > n$ that is true for any n. Therefore $\forall x \exists y A(x, y)$ is true under the given interpretation.

Let us consider next the formula $\exists y \forall x A(x, y)$ under the same interpretation. As before, $\exists y \forall x\, y > x$ is true if there is some m such that $\forall x\, m > x$ is true. The latter is true if, whatever value is given to n, $m > n$ is true. If we give to n the value m, we get $m > m$ that is false. Therefore $\exists y \forall x A(x, y)$ is false under the given interpretation.

Let us next combine the two formulas into the implication $\forall x \exists y A(x, y) \supset \exists y \forall x A(x, y)$. Under the given interpretation, the antecedent is true but the consequent false, so that the implication $\forall x \exists y A(x, y) \supset \exists y \forall x A(x, y)$ is false under the interpretation. We say that the formula $\forall x \exists y A(x, y) \supset \exists y \forall x A(x, y)$ is **refutable** and that there is an **arithmetic counterexample** to it.

A formula is **closed** if it has no free variables. **Logical truth** of a closed formula A in classical predicate logic requires that A be true under any interpretation, i.e., true whatever domain of individuals $\mathcal{D}$ is chosen and whatever concrete meaning is given to the schematic atomic formulas. If a formula A contains the free variables $x, \dots, z$, logical truth is taken to refer to the logical truth of its **universal closure** $\forall x \dots \forall z A$.

Classical predicate logic is often explained as follows: To show a formula A logically true, one tries to disprove it by a falsifying counterexample. If such a counterexample turns out impossible, A is logically true. Another way to express the matter is: Define first a formula to be **satisfiable** if it is true under at least one interpretation. The search for a counterexample to formula A can now be described as the search for an interpretation that shows $\neg A$ satisfiable. The connection to refutability is: A is refutable whenever $\neg A$ is satisfiable, and A is satisfiable whenever $\neg A$ is refutable.

When it is assumed that atomic formulas have determinate truth values under an interpretation, it is required that they contain just constants. We determined some such truth values above, e.g., we noticed that $n + 1 > n$ is true whatever constant n is chosen, and that $m > m$ is similarly false. These truth values are determined on the basis of the mathematical properties of the domain $\mathcal{N}$.

There need not exist any general method for determining the truth value of an atomic formula under an interpretation. As an example, let the domain consist of the real numbers $\mathcal{R}$, and consider the relation of equality $x = y$. Let C be the well-known Riemann constant. A decimal expansion for C can be computed to any length. So far the computation has given the value $0.4999\dots$ to millions of decimals, but nobody knows if a decimal less than 9 will show up. If not, $2C = 0.999\ldots = 1$, otherwise $2C < 1$. Thus, nobody knows if the atomic formula $2C = 1$ is true in $\mathcal{R}$. If it is decided one day, other undecided examples can be pointed out.

As a further example, consider the converse of the above implication: $\exists y \forall x A(x, y) \supset \forall x \exists y A(x, y)$. We show that it is true under any interpretation. For this, it is sufficient to take an arbitrary interpretation, to assume that $\exists y \forall x A(x, y)$ is true under this interpretation, then to show that also $\forall x \exists A(x, y)$ is true under the interpretation. Henceforth, we shall not repeat the words 'under the interpretation'. If $\exists y \forall x A(x, y)$ is true, there is some b such that $\forall x A(x, b)$ is true. Then $A(a, b)$ is true for any a. Therefore $\exists y A(a, y)$ is true for any a. Therefore $\forall x \exists A(x, y)$ is true, and so is the implication $\exists y \forall x A(x, y) \supset \forall x \exists y A(x, y)$. The interpretation was arbitrary, and therefore $\exists y \forall x A(x, y) \supset \forall x \exists y A(x, y)$ is logically true.

The above proof of logical truth uses principles of proof on a semantical level that are absolutely analogous to those we learned when studying the rules of inference for the quantifiers. An informal logical proof of the implication would be more or less as follows: Assume $\exists y \forall x A(x, y)$. Next, to apply rule $\exists E$, consider an instance $\forall x A(x, y)$ with an eigenvariable y. From $\forall x A(x, y)$, conclude $A(x, y)$ by $\forall E$, then $\exists y A(x, y)$ by $\exists I$. Now the temporary assumption $\forall x A(x, y)$ in rule $\exists E$ can be closed. Noting that x works as an eigenvariable in rule $\forall I$, conclude $\forall x \exists y A(x, y)$, and as a last step, close the open assumption to conclude by rule $\supset I$ $\exists y \forall x A(x, y) \supset \forall x \exists y A(x, y)$.

It has often been said that the semantics of classical predicate logic just repeats the principles of logic reasoning on an informal, semantical level, instead of explaining them. Let us try to elaborate on this general remark.

All formulas in classical predicate logic have an equivalent in prenex normal form, and say we have the formula $\forall x \exists y \forall z A(x, y, z)$. To determine whether it is logically true, consider an arbitrary domain, and try to find a falsifying counterexample to it. Take an instance $\exists y \forall z A(a, y, z)$, then some b to obtain $\forall z A(a, b, z)$. Finally, take an instance $A(a, b, c)$. A falsifying instance results if this propositional formula is not a tautology, something that can be in principle checked by a truth table. The crux of the matter is, of course, to choose a and c in an appropriate way to arrive at such a falsifying instance, or to show that no such instance can be found. The latter would make $\forall x \exists y \forall z A(x, y, z)$ a logical truth.

The explanation of the connectives and quantifiers in terms of proof proceeded through the introduction rules: These rules gave sufficient conditions under which a corresponding formula could be inferred. If the semantics of classical predicate logic is described in terms of an attempt at finding a counterexample to a formula in prenex normal form, the reasoning on the semantical level uses the elimination rules for the quantifiers, as is seen from the example of the preceding paragraph.

From the point of view of semantics, it is the elimination rules that give meaning, whereas from the point of view of proof systems, it is the introduction rules that have this task.

Syntax builds up, semantics unfolds, we could put briefly.

(b) Kripke semantics of intuitionistic predicate logic. We shall just outline the basic ideas. In place of a fixed domain, the definition of a Kripke model for predicate logic has a nonempty domain of objects for each world w, denoted $\mathcal{D}_w$. New objects are always added as one moves along the accessibility relation, but never taken away:

Monotonicity of the domain function. *If a is an object in $\mathcal{D}_w$ and $w \leqslant o$, then a is an object in $\mathcal{D}_o$.*

The forcing relation of Table 7.4 is extended to the quantifiers as follows:

Table 10.1 Forcing the quantifiers

6. $w \Vdash \forall x A(x)$ whenever from $w \leqslant o$ follows $o \Vdash A(a)$ for all a in $\mathcal{D}_o$.
7. $w \Vdash \exists x A(x)$ whenever $w \Vdash A(a)$ for some a in $\mathcal{D}_w$.

The difference to the classical explanation is not about existence, in clause 7, but in clause 6 for universality: Any object in any possible future world has to be considered in that clause.

10.2 Completeness

The most important result about classical predicate logic is:

Completeness of classical predicate logic. *For each closed formula A in classical predicate logic, either A is derivable or A has a counterexample.*

A proof can be found in *Structural Proof Theory*, section 4.4.

In a failed proof search that does not terminate, rules $R\exists$ and $L\forall$ generate ever new terms. The completeness result means that such failed proof search can be turned, after a finite number of steps, into a counterexample. However, it is not possible to determine from a given formula A in advance any upper bound on instantiations in proof search. Were this possible, the systematic search for a counterexample could be turned into a decision method. As remarked at the end of the previous chapter, such is not the case:

Undecidability of classical predicate logic. *There is no algorithm for deciding if a formula is derivable in classical predicate logic.*

There is, for each formula A of classical predicate logic, a classically equivalent formula A^* such that if A is a theorem of classical predicate logic, A^* is a theorem of intuitionistic predicate logic. Therefore, if the latter possessed a decision method, also the former would, contrary to what was stated.

Finally, we have the result established by Skolem in his talk in Helsingfors in 1922:

Löwenheim–Skolem theorem. *If a formula A is satisfiable, it is true under some arithmetic interpretation.*

One way to express the completeness of classical predicate logic is: If A is logically true, it is derivable. To see this, assume A to be logically true. Then it is true under any interpretation and cannot have a counterexample. By the completeness result, it is derivable.

From a metaphysical point of view, the completeness of classical predicate logic seems at first an immense result. Any talk about any objects, their properties and relations, universality and existence, can be codified in a logical calculus that has a finite, indeed quite low number of rules, such that any logical truth about these objects and their properties can be established by these rules and communicated by a finite message to anyone who knows the rules. It is, *a priori*, not clear that this should be the case. The Löwenheim–Skolem theorem, however, tones down somewhat the importance of completeness. One way to look at the theorem is that the language of predicate logic is not able to distinguish between natural numbers and other objects: Any consistent collection of formulas in predicate logic has an arithmetic interpretation.

10.3 Interpretation of classical logic in intuitionistic logic

At the end of Section 5.2, it was noted that a formula A is derivable in classical propositional logic if and only if $\neg\neg A$ is derivable in intuitionistic propositional logic. The same is not true of predicate logic. An example is the formula $\forall x(A(x) \vee \neg A(x))$ that is classically derivable. It is somewhat strange that the double-negated formula $\neg\neg\forall x(A(x) \vee \neg A(x))$ is not intuitionistically derivable. A more complicated transformation can be defined, denoted by $(A)^*$, and it has an effect analogous to the addition of double-negations in the head of formulas in propositional logic:

Table 10.2 The Gödel–Gentzen translation

1.	P^*	$\rightsquigarrow$	$\neg\neg P$
2.	$(\neg A)^*$	$\rightsquigarrow$	$\neg(A^*)$
3.	$(A \,\&\, B)^*$	$\rightsquigarrow$	$(A^*) \,\&\, (B^*)$
4.	$(A \supset B)^*$	$\rightsquigarrow$	$(A^*) \supset (B^*)$
5.	$(\forall x A(x))^*$	$\rightsquigarrow$	$\forall x(A(x)^*)$
6.	$(A \vee B)^*$	$\rightsquigarrow$	$\neg(\neg(A^*) \,\&\, \neg(B^*))$
7.	$(\exists x A(x))^*$	$\rightsquigarrow$	$\neg\forall x\neg(A(x)^*)$

It is obvious that A and $(A)^*$ are classically equivalent.

Gödel's translation removes also implications $A \supset B$, by translating them into $\neg((A^*) \,\&\, \neg(B^*))$.

The crucial points of the translation are disjunction and existence. The former is translated into the intuitionistically weaker $\neg(\neg(A^*) \,\&\, \neg(B^*))$, and the latter into $\neg\forall x\neg(A^*)$. Gödel's and Gentzen's translations were made for interpreting classical arithmetic in intuitionistic arithmetic. The atomic formulas are decidable equations and need therefore not be double-negated, but Gentzen notes that if it is done, as in Table 10.2, an interpretation of classical predicate logic in intuitionistic predicate logic follows:

Translation from classical to intuitionistic logic. *If A is derivable in classical predicate logic, then A^* is derivable in intuitionistic predicate logic.*

Before Gödel and Gentzen found the translations, it was thought that the indirect existence proofs of classical arithmetic would be in need of a separate foundation through a proof of consistency. Gödel's and Gentzen's result shows that this is not the case: If a contradiction is derivable classically, it is already derivable intuitionistically, i.e., without indirect proofs.

PART III

Beyond pure logic

11 | Equality and axiomatic theories

11.1 Equality relations

(a) The axioms of equality. We assume given a domain $\mathcal{D}$ of individuals $a, b, c \ldots$ and a two-place relation $a = b$ in $\mathcal{D}$ with the following standard axioms:

Table 11.1 The axioms of an equality relation

EQ1.	*Reflexivity:*	$a = a$
EQ2.	*Symmetry:*	$a = b \supset b = a$
EQ3.	*Transitivity:*	$a = b \;\&\; b = c \supset a = c$

These axioms can be added to a Frege-style axiomatization of logic. We shall instead first add them to natural deduction, so that instances of the axioms can begin a derivation branch. Thus, when we ask whether a formula A is derivable from the collection of assumption formulas Γ by the axioms of equality, arbitrary instances of the axioms can be added to Γ.

Let us take as an example a derivation of $d = a$ from the assumptions $a = b$, $c = b$, and $c = d$:

$$\dfrac{a=d \supset d=a \qquad \dfrac{a=c \,\&\, c=d \supset a=d \qquad \dfrac{\dfrac{a=b \,\&\, b=c \supset a=c \qquad \dfrac{a=b \qquad \dfrac{c=b \supset b=c \qquad c=b}{b=c}{\scriptstyle \supset E}}{a=b \,\&\, b=c}{\scriptstyle \&I}}{a=c}{\scriptstyle \supset E} \qquad c=d}{a=c \,\&\, c=d}{\scriptstyle \&I}}{a=d}{\scriptstyle \supset E}}{d=a}{\scriptstyle \supset E}$$

Each top formula in the derivation is either one of the assumptions or an instance of an equality axiom. The derivation tree looks somewhat forbidding. The natural way to reason would be different, something like: from a to b, b to c, c to d, therefore from d to a. Here the principles are that equalities can be combined in chains and that equalities go both ways. The latter was applied to get the link b to c from c to b and to get the conclusion d to a from a to d.

Logic in the derivation of $d = a$ from the assumptions $a = b$, $c = b$, and $c = d$ seems like a decoration necessitated by the use of logic in the writing of the axioms. We now want to say instead that two equalities $a = b$ and $b = c$ give at once $a = c$, and that $a = b$ gives $b = a$, i.e., we reformulate the axioms as **rules of inference**. The procedure is analogous to that of Gentzen who reformulated the axiomatic logic of Frege and others as the rule system of natural deduction.

Table 11.2 Symmetry and transitivity as rules of inference

$$\frac{a = b}{b = a}\,Sym \qquad \frac{a = b \quad b = c}{a = c}\,Tr$$

Our example derivation becomes:

$$\dfrac{\dfrac{\dfrac{a = b \quad \dfrac{c = b}{b = c}\,Sym}{a = c}\,Tr \qquad c = d}{a = d}\,Tr}{d = a}\,Sym$$

This transparent rule-based derivation should be compared with the axiomatic derivation above. To get the full theory of equality, we must add reflexivity as a zero-premiss rule:

Table 11.3 The rule of reflexivity

$$\overline{a = a}\,Ref$$

Now formal derivations start from assumptions and instances of rule *Ref.*

What about the role of logic after the addition of mathematical axioms as rules? A premiss of an equality rule can be the conclusion of a logical rule and a conclusion of an equality rule a premiss in a logical rule. It should be clear that logic itself must not be 'creative' in the sense of making equations derivable from given equations used as assumptions, if they were not already derivable by just the equality rules. To show that there cannot be any such creative use of logic, Gentzen's normalization theorem comes to help. No introduction rule can have a premiss of a mathematical rule as a conclusion, because the latter are atomic and don't have any logical structure. By the same reason, no conclusion of a mathematical rule can be the major premiss of an E-rule. Therefore the mathematical rules can be completely separated from the logical ones, so that in a normal derivation, the former are applied first in a part of derivation that contains only atomic formulas, then the latter build up logical structure. Thus, if an equality is derivable from given equalities in natural deduction extended with the rules of equality, it is derivable by just the rules of equality.

The separation of logic from mathematical axioms goes through for a large class of axiomatizations, though not all. In particular, axioms that contain just free variables but no quantifiers can be converted to rules that maintain the separation.

Assume there to be a derivation of the equality $a = c$ from given assumptions $a_1 = c_1, \ldots, a_n = c_n$ by the rules of equality. By what has been said, no logical rules need be used. Assume there to be a term b in the derivation that is neither a term in the conclusion $a = c$ nor a term in any of the assumptions. There is thus some instance of rule *Tr* that removes b:

$$\frac{a = b \quad b = c}{a = c}\,Tr$$

If the premiss $a = b$ is a conclusion of rule *Tr*, we can permute up the instance of *Tr* that removes b, as follows:

$$\frac{\dfrac{a = d \quad d = b}{a = b}\,Tr \qquad b = c}{a = c}\,Tr \qquad\qquad \frac{a = d \qquad \dfrac{d = b \quad b = c}{d = c}\,Tr}{a = c}\,Tr$$

A similar transformation applies if the second premiss $b = c$ has been derived by *Tr*. Thus, we may assume that neither premiss of the step of *Tr* that removes the 'unknown' term b has been derived by *Tr*. The next possibility is that both premisses have been derived by rule *Sym*. We then have the part of derivation and its transformation:

$$\frac{\dfrac{b = a}{a = b}\,Sym \qquad \dfrac{c = b}{b = c}\,Sym}{a = c}\,Tr \qquad\qquad \dfrac{\dfrac{c = b \quad b = a}{c = a}\,Tr}{a = c}\,Sym$$

In the end, at least one premiss of the step of *Tr* that removes the term b has an instance of rule *Ref* as one premiss, as in:

$$\frac{d = b \qquad \dfrac{}{b = b}\,Ref}{d = b}\,Tr$$

Now the conclusion is equal to the other premiss, so the step of *Tr* can be deleted. Tracing up in the derivation the premiss $d = b$, the permutations can never lead to an instance of *Tr* that removes b and has an assumption as one premiss, because then b would be a term known from the assumption. The conclusion is that there cannot be derivations by the rules of equality with unknown terms.

Consider next a derivation that has a 'cycle' or a 'loop', i.e., a branch in which the same equality occurs twice:

$$\begin{array}{c} \vdots \\ a = b \\ \vdots \\ a = b \\ \vdots \end{array}$$

The part between the two occurrences can be cut out. This part may use some equalities as assumptions that are not otherwise used in the derivation, but their deletion just improves the result: The conclusion follows from fewer assumptions. When no loops are allowed to occur, all derivations of an equality $a = c$ from the assumptions $a_1 = c_1, \ldots, a_n = c_n$ have an upper bound on size, here the length of the longest derivation tree branch: The number of distinct terms is at most $2n + 2$, therefore the number of distinct equalities is at most $(2n + 2)^2$, an upper bound on height.

The role of reflexivity in derivations is to conclude a reflexivity atom. In other cases, reflexivity will produce a loop. As for transitivity and symmetry, their role is now clear. We can conclude $a_1 = a_n$ from assumptions $a_1 = a_2, \ldots a_{n-1} = a_n$ that form, possibly after some equalities have been reversed by *Sym*, a chain.

(b) Natural deduction for predicate logic with equality. Predicate logic with equality is obtained from standard predicate logic through the addition of a two-place reflexive relation $a = b$ with the property that equals be substitutable everywhere. The latter is formulated as:

The replacement axiom. $A(a)\ \&\ a = b \supset A(b)$

It is possible to restrict the replacement axiom to atomic predicates and relations. Therefore it is also possible to consider predicate logic with equality as a system of natural deduction extended by two mathematical rules that operate on atomic formulas:

Table 11.4 The rules of predicate logic with equality

$$\frac{}{a = a}\,\textit{Ref} \qquad \frac{P(a) \quad a = b}{P(b)}\,\textit{Rep}$$

The second rule is schematic: There is one rule for each predicate and relation. The normal form of derivations, as explained in Section 3.4, carries over to predicate calculus with equality. The reason is that the rules of equality act on atomic formulas that cannot be major premisses of elimination rules. Thus, the normalization process by which derivations with derived major premisses of E-rules are converted into ones in which they are assumptions, is not affected by the presence of instances of equality rules. These latter can be permuted above the logical rules.

We show first that the equality of predicate logic with equality is an equality relation:

Lemma 11.1. *Rules* Sym *and* Tr *are derivable in predicate logic with equality.*

Proof. For *Sym*, set $P(x) \equiv x = a$ in the rule of replacement. The conclusion of *Sym* is derived from its premiss as follows:

$$\frac{\dfrac{}{a = a}\,\textit{Ref} \qquad a = b}{b = a}\,\textit{Rep}$$

For *Tr*, set $P(x) \equiv x = c$. The conclusion of *Tr* is derived from its premisses by:

$$\frac{b = c \qquad \dfrac{a = b}{b = a}\,\textit{Sym}}{a = c}\,\textit{Rep}$$

QED.

The lemma shows, incidentally, that the theory of equality of part (a) above is a special case of predicate logic with equality, namely the one in which there are no other predicates than equality.

We show that the rule of replacement of Table 11.4 is admissible for arbitrary formulas:

Lemma 11.2. *Application of replacement to arbitrary formulas reduces to rule* Rep.

Proof. The proof is by induction on the length of the replacement formula. The base case is that of an atomic formula, covered by rule *Rep*. For $\bot$, nothing happens. The other cases are:

1. The formula is $A(a)\,\&\,B(a)$. Replacement is reduced to the components $A(a)$ and $B(a)$ as follows:

$$\dfrac{A(a)\,\&\,B(a) \qquad \dfrac{\dfrac{\overset{1}{A(a)} \quad a=b}{A(b)}{\scriptstyle Rep} \qquad \dfrac{\overset{1}{B(a)} \quad a=b}{B(b)}{\scriptstyle Rep}}{A(b)\,\&\,B(b)}{\scriptstyle \&I}}{A(b)\,\&\,B(b)}{\scriptstyle \&E,1}$$

2. With $A(a) \vee B(a)$, the reduction is similar.

3. With $A(a) \supset B(a)$, the reduction is as follows:

$$\dfrac{\dfrac{A(a) \supset B(a) \qquad \dfrac{\overset{2}{A(b)} \quad \dfrac{a=b}{b=a}{\scriptstyle Sym}}{A(a)}{\scriptstyle Rep} \qquad \dfrac{\overset{1}{B(a)} \quad a=b}{B(b)}{\scriptstyle Rep}}{B(b)}{\scriptstyle \supset E,1}}{A(b) \supset B(b)}{\scriptstyle \supset I,2}$$

The quantifiers are treated similarly. QED.

By this proof, the calculus is complete. Notice that if the standard implication elimination rule were used, the replacement of a with b in $B(a)$ would have to be done after the logical elimination step so that logical and mathematical parts of derivations could not be maintained apart.

We can give a complete analysis of the role of rule *Ref* in derivations. If *Ref* gives the second premiss of rule *Rep*, a loop is produced. If it gives the first premiss, the replacement predicate is of the form $a = x$ or $x = a$. There are thus altogether three possible steps:

$$\dfrac{P(a) \quad \dfrac{}{a=a}{\scriptstyle Ref}}{P(a)}{\scriptstyle Rep} \qquad \dfrac{\dfrac{}{a=a}{\scriptstyle Ref} \quad a=b}{a=b}{\scriptstyle Rep} \qquad \dfrac{\dfrac{}{a=a}{\scriptstyle Ref} \quad a=b}{b=a}{\scriptstyle Rep}$$

Two cases give a loop, the third the rule of symmetry.

(c) Sequent calculus for predicate logic with equality. The rules of equality were formulated in the previous subsection for a system of intuitionistic natural deduction. They can be equally well formulated for sequent calculus, as is natural if we want to have a system of classical predicate logic with equality. Our specific aim will be to show that if the sequent $\Gamma \vdash \Delta$ is derivable and contains no equality, the rules for equality are not needed.

In other words, predicate logic with equality is conservative over predicate logic without equality.

It is usual in sequent calculus to let derivations start with 'axiomatic sequents' of the form:

Table 11.5 Replacement rules as axiomatic sequents

$\vdash a = a$	$a = b,\ P(a) \vdash P(b)$

Here P is an atomic formula. The use of axiomatic sequents requires an addition to sequent calculus, namely the **rule of cut**. This rule is explained in Section 13.4. It can be avoided if the replacment axioms are converted into rules instead of axiomatic sequents:

Table 11.6 Replacement rules in sequent calculus

$\dfrac{a = a, \Gamma \vdash \Delta}{\Gamma \vdash \Delta}\,Ref$	$\dfrac{a = b,\ P(a),\ P(b), \Gamma \vdash \Delta}{a = b,\ P(a), \Gamma \vdash \Delta}\,Rep$

The axiomatic sequents follow at once by the rules:

$$\frac{a = a \vdash a = a}{\vdash a = a}\,Ref \qquad \frac{a = b,\ P(a),\ P(b) \vdash P(b)}{a = b,\ P(a) \vdash P(b)}\,Rep$$

In the other direction, the conclusions of the rules can be derived from their premisses by axiomatic sequents and cuts.

Lemma 11.3. *The replacement rule*

$$\frac{a = b,\ A(a),\ A(b), \Gamma \vdash \Delta}{a = b,\ A(a), \Gamma \vdash \Delta}\,Rep$$

is admissible for arbitrary formulas A.

Proof. The proof is a sequent calculus version of the proof given in lemma 11.2. QED.

Next we prove the conservativity of predicate logic with equality over predicate logic. Assume there to be a derivation of a sequent $\Gamma \vdash \Delta$ that has no equalities. We show that rules *Ref* and *Rep* are not needed in the derivation. The only way to arrive at such a sequent $\Gamma \vdash \Delta$ from sequents that contain equalities is by rule *Ref*. Thus, it is sufficient to show that instances *Ref* can be eliminated from derivations of equality-free sequents.

The rule of replacement has an instance that produces a **duplication** of an atom, namely, when the predicate $P(x)$ is $x = b$:

$$\frac{a = b,\, a = b,\, b = b,\, \Gamma \vdash \Delta}{a = b,\, a = b,\, \Gamma \vdash \Delta}\,Repl$$

The conclusion follows by rule *Ref*, but we want to show that that rule need not be used. Therefore, for the proof of conservativity, the instance of *Rep* with the duplication of $a = b$ **contracted** has to be added to the rules:

$$\frac{a = b,\, b = b,\, \Gamma \vdash \Delta}{a = b,\, \Gamma \vdash \Delta}\,Rep^*$$

The rules can remove equalities from the antecedent part of sequents, but not from the succedent part. Therefore we have the result:

Lemma 11.4. *If* $\Gamma \vdash \Delta$ *has no equalities and is derivable in classical sequent calculus extended by the rules* Ref+Rep+Rep*, *no sequents in its derivation have equalities in the succedent.*

The following lemma contains the essential analysis in the proof of conservativity of predicate logic with equality over predicate logic:

Lemma 11.5. *If* $\Gamma \vdash \Delta$ *has no equalities and is derivable through the use of the rules* Ref+Rep+Rep*, *it is derivable by* Rep *and* Rep*.

Proof. We show that all instances of *Ref* can be eliminated from a given derivation, by induction on the height of derivation of a topmost instance

$$\frac{a = a,\, \Gamma' \vdash \Delta'}{\Gamma' \vdash \Delta'}\,Ref$$

If the premiss is an initial sequent, also the conclusion is, because by the above lemma, Δ' contains no equality. If the premiss has been concluded by a logical rule, apply the inductive hypothesis to the premisses and then the rule.

If the premiss has been concluded by *Rep* there are two cases, according to whether $a = a$ is or is not principal. In the latter case the derivation is, with $\Gamma' = P(b), \Gamma''$,

$$\cfrac{\cfrac{a = a,\, b = c,\, P(b),\, P(c),\, \Gamma'' \vdash \Delta'}{a = a,\, b = c,\, P(b),\, \Gamma'' \vdash \Delta'}\,Rep}{b = c,\, P(b),\, \Gamma'' \vdash \Delta'}\,Ref$$

By permuting the two rules, the inductive hypothesis can be applied.

If $a = a$ is principal, the derivation is, with $\Gamma' \equiv P(a), \Gamma''$,

$$\cfrac{\cfrac{a = a,\ P(a),\ P(a), \Gamma'' \vdash \Delta'}{a = a,\ P(a), \Gamma'' \vdash \Delta'}\,\mathit{Repl}}{P(a), \Gamma'' \vdash \Delta'}\,\mathit{Ref}$$

It can be shown (cf. *Structural Proof Theory*, p. 131) that the two occurrences of $P(a)$ can be contracted to one without increasing the height of derivation. Therefore there is a derivation of $a = a,\ P(a), \Gamma'' \vdash \Delta'$ to which the inductive hypothesis applies, giving a derivation of $\Gamma' \vdash \Delta'$ without rule *Ref*.

If the premiss of *Ref* has been concluded by Rep^* with $a = a$ not principal the derivation is:

$$\cfrac{\cfrac{a = a, b = c, c = c, \Gamma' \vdash \Delta'}{a = a, b = c, \Gamma' \vdash \Delta'}\,\mathit{Rep^*}}{b = c, \Gamma'' \vdash \Delta'}\,\mathit{Ref}$$

The rules are permuted and the inductive hypothesis applied.

If $a = a$ is principal the derivation is:

$$\cfrac{\cfrac{a = a, a = a, \Gamma' \vdash \Delta'}{a = a, \Gamma' \vdash \Delta'}\,\mathit{Rep^*}}{\Gamma' \vdash \Delta'}\,\mathit{Ref}$$

Now apply height-preserving contraction and the inductive hypothesis.

QED.

Next, because the rules *Rep* and Rep^* have equalities in their conclusions, we obtain:

Theorem 11.6. *If* $\Gamma \vdash \Delta$ *is derivable in classical sequent calculus extended by the rules* Ref+Rep+Rep* *and if* Γ, Δ *contains no equality, then* $\Gamma \vdash \Delta$ *is derivable without these rules.*

11.2 Sense and denotation

One reads often of predicate logic with **identity**, instead of predicate logic with equality. There have been many attempts in philosophy at defining identity. Leibniz had the idea that two objects are identical if they share the same properties. Such a definition can be written in **second-order logic** in which it is allowable to quantify over predicates. Upper case variables $X, Y, Z, \ldots$ are used for predicates. The definition of identity is:

$$a = b \equiv \forall X(X(a) \supset\subset X(b))$$

It is obvious that identity defined in this way has the properties of an equality relation, as in Table 11.1. Moreover, each instance of the schematic replacement axiom $A(a) \,\&\, a = b \supset A(b)$ follows from the definition.

Logical systems, their axioms and rules, operate on expressions for objects, not the objects themselves. The former **denote** the latter. The same is true of such expressions as $7 + 5$ and 3×4. They are different expressions for the same object. In the equality $7 + 5 = 3 \times 4$, we have a sum at left and a product at right. The equality of $7 + 5$ and 3×4 means that these different expressions **have the same value**, i.e., denote the same natural number 12.

The **sense** of an expression is the way it has been built, its **denotation** or **reference** is the object expressed. Thus, the sense of $7 + 5$ is that it is the sum of two numbers. Another terminology is **intension** and **extension**. A simple example may be useful. Consider plane elementary geometry as it has been taught at school. We have two kinds of objects, namely points $a, b, c, \ldots$ and lines $l, m, n, \ldots$. There are relations of equality for points $a = b$ and for lines $l = m$, a relation of **parallelism** $l \parallel m$, and a relation of **incidence** written $a \in l$ and read *point a is incident with line l.* Finally, given a point a and a line l, we can construct the object $par(l, a)$, *the parallel to line l through point a.*

The **axioms of geometry** include equality axioms for the equality of points and of lines, as in Table 11.1, and replacement axioms for equals in the parallelism and incidence relations. The replacement axioms are:

$$l \parallel m \,\&\, m = n \supset l \parallel n$$
$$a \in l \,\&\, a = b \supset b \in l$$
$$a \in l \,\&\, l = m \supset a \in m$$

The relation of parallelism is reflexive, symmetric, and transitive. Thus, it is an equality relation. It is, on the other hand, clear that two parallel lines are not in general equal, because they can be in different places in the geometrical plane. The terminology **equivalence relation** is often used in such situations.

Two objects can be equivalent as far as the properties and relations considered are concerned or as one says, **extensionally equal**, even though they need not be the same in any ultimate sense. Such extensional equality is precisely what replacement axioms correspond to.

The famous **axiom of parallels** can be written with the help of predicate logic in the following way:

$$l \parallel m \,\&\, a \in l \,\&\, a \in m \supset l = m$$

In words, *any two lines parallel to each other that have a common point are equal to each other.*

Consider the objects $par(l, a)$ and $par(m, a)$. These are **constructed objects**, and there is no limit to how complicated such constructed objects can be. There would normally be in geometry a **connecting line** construction $ln(a, b)$ and an **intersection point** construction $pt(l, m)$, and we can iterate constructions to get, say:

$$ln(pt(par(l, a), m), b)$$

First the parallel to l through a has been drawn, then its intersection point with m, and finally the connecting line of this point and b has been drawn.

As said, the sense of an object is the way it has been constructed. Constructed objects in geometry have certain ideal properties. The parallel line construction has the following characteristic ideal properties, expressed as axioms for the construction par:

$$a \in par(l, a) \qquad par(l, a) \parallel l$$

The parallel goes exactly through the point a and is exactly parallel to the given line l. We can draw geometric figures that are far from being ideal objects, even by free hand, because we understand what it means for them to have the ideal properties and can reason about the figures as if they had those properties.

Given two lines l and m and a point a, we can thus construct the lines $par(l, a)$ and $par(m, a)$ and know by the construction axioms that $a \in par(l, a)$ and $a \in par(m, a)$. In other words, lines l and m pass through a common point a. If at some stage we find that the two given lines are parallel, $l \parallel m$, we can conclude that $par(l, a) \parallel par(m, a)$ and by the axiom of parallels that these latter are equal, through the instance:

$$\begin{aligned}&par(l, a) \parallel par(m, a) \,\&\, a \in par(l, a) \,\&\, a \in par(m, a) \supset \\ &\quad par(l, a) = par(m, a)\end{aligned}$$

The two lines $par(l, a)$ and $par(m, a)$ are equal in the sense that they are in the same place, in geometrical terminology **congruent**, but they are not identical, whatever that means, because they have been constructed in different ways. Even though identity is an elusive notion, it should include that identity is immediately recognizable, but that need not be the case for equality in the geometric example: The equality $par(l, a) = par(m, a)$ was concluded from $l \parallel m$, but that parallelism may be as difficult to establish as anything.

11.3 Axiomatic theories

Predicate logic can be used for the formulation of many mathematical theories, namely those that correspond to its natural ontology of individual objects, their properties, and relations between the objects. Such theories include elementary arithmetic, geometry, and algebra, the second one of which was illustrated at the end of the previous section. Other mathematical theories use concepts such as arbitrary sets of objects from a given collection that go beyond the expressive means of predicate logic. For example, real numbers can be construed as sets of natural numbers, advanced parts of geometry can use continuity axioms, and advanced parts of algebra principles from set theory.

The theory of equality can be considered a model of a simple axiomatic theory formulated in predicate logic. We shall treat one less simple theory as an example, namely lattice theory. Its origins are in great part in logic itself, in the algebraic tradition that prevailed in logic from the times of George Boole around 1850 through Ernst Schröder, Charles Peirce, and others, to Skolem's work around 1920 (cf. also Section 14.2). After this time, lattice theory became a part of algebra.

A **lattice** consists of domain $\mathcal{D}$ of objects $a, b, c, \ldots$ with a relation of **partial order** $a \leqslant b$. An equality relation is defined by $a = b \equiv a \leqslant b \,\&\, b \leqslant a$. Next we have two operations by which new objects can be constructed:

$a \wedge b$ the **meet** of a and b
$a \vee b$ the **join** of a and b

The axioms are:

I General properties of the basic relation

Reflexivity: $a \leqslant a$ *Transitivity:* $a \leqslant b \,\&\, b \leqslant c \supset a \leqslant c$

II Properties of constructed objects

$$a \wedge b \leqslant a \quad a \wedge b \leqslant b \quad a \leqslant a \vee b \quad b \leqslant a \vee b$$

III Uniqueness of constructed objects

$$c \leqslant a \,\&\, c \leqslant b \supset c \leqslant a \wedge b$$
$$a \leqslant c \,\&\, b \leqslant c \supset a \vee b \leqslant c$$

By axiom III, anything in the partial order that is 'between' $a \wedge b$ and a and also between $a \wedge$ b and b, is equal to $a \wedge b$:

$$a \wedge b \leqslant c \,\&\, c \leqslant a \,\&\, c \leqslant b \supset c = a \wedge b$$

The substitution principles for equals in the meet and join operations are:

$$a = b \supset a \wedge c = b \wedge c \qquad b = c \supset a \wedge b = a \wedge c$$

These are provable.

Part of the origin of lattice theory is in number theory: Richard Dedekind noticed that the greatest common divisor and least common multiple of two natural numbers follow certain abstract laws, namely those for a lattice meet and join. Lattice theory was practised by Ernst Schröder in his 'algebra of logic', though with a terminology and notation that is completely different from that of today. Schröder considered the theory quite abstractly, with various readings of the lattice order relation $a \leqslant b$: The most common reading was that a and b were some sort of domains and the order an inclusion relation, so, in substance, sets with a subset relation. Then meet and join became intersection and union, respectively. In another reading, a and b could be taken as propositions and the order expressed logical consequence with meet and join standing for conjunction and disjunction, or they could be taken as 'circumstances' with a relation of cause and effect. Skolem's early work in logic followed Schröder's algebraic tradition. In 1920, he solved what is called today the word problem for freely generated lattices. Schröder's terminology and notation were unknown to the extent that Skolem's discovery remained unnoticed until 1992.

A peculiarity of Skolem's axiomatization is that it does not use the lattice operations, but additional basic relations. Why he made this axiomatization is not told to the reader of his article, but it works as a fine illustration of a **relational axiomatization.**

Relational axiomatizations replace operations with relations. For example, relational lattice theory is based on the idea of adding two three-place relations $M(a, b, c)$, $J(a, b, c)$, read as 'the meet of a and b is c' and 'the join of a and b is c', and axioms that state the existence of meets and joins:

$$\forall x \forall y \exists z M(x, y, z) \quad \forall x \forall y \exists z J(x, y, z)$$

There are many more axioms as compared to an axiomatization with operations, but there are no functions:

I General properties of the basic relations

Reflexivity: $a \leqslant a$ *Transitivity:* $a \leqslant b \,\&\, b \leqslant c \supset a \leqslant c$

II Properties of meet and join

$$M(a, b, c) \supset c \leqslant a \quad M(a, b, c) \supset c \leqslant b$$
$$J(a, b, c) \supset a \leqslant c \quad J(a, b, c) \supset b \leqslant c$$

III Uniqueness of meet and join

$$M(a, b, c) \,\&\, d \leqslant a \,\&\, d \leqslant b \supset d \leqslant c$$
$$J(a, b, c) \,\&\, a \leqslant d \,\&\, b \leqslant d \supset c \leqslant d$$

Substitution of equals in the meet and join relations needs to be postulated, with $a = b \equiv a \leqslant b \,\&\, b \leqslant a$. The relations have three arguments, so to cut down the number of axioms, we do as follows:

IV Substitution axioms

$$M(a, b, c) \,\&\, a = d \,\&\, b = e \,\&\, c = f \supset M(d, e, f)$$
$$J(a, b, c) \,\&\, a = d \,\&\, b = e \,\&\, c = f \supset J(d, e, f)$$

Substitution in one argument, say d for a in $M(a, b, c)$, is obtained by the instance $M(a, b, c) \,\&\, a = d \,\&\, b = b \,\&\, c = c \supset M(d, b, c)$.

Finally, we have:

V Existence of meets and joins

$$\forall x \forall y \exists z M(x, y, z) \quad \forall x \forall y \exists z J(x, y, z)$$

The existential axioms are used as follows:

$$\cfrac{\cfrac{\forall x \forall y \exists z M(x, y, z)}{\exists z M(a, b, z)}\,{\scriptstyle \forall E, \forall E} \qquad \begin{matrix} \overset{1}{M(a, b, v)} \\ \vdots \\ C \end{matrix}}{C}\,{\scriptstyle \exists E, 1}$$

Here v is an eigenvariable of rule $\exists E$. One would normally use existential axioms by simply considering an instance $M(a, b, v)$ with v arbitrary. This was done by Skolem in 1920, well before Gentzen gave the natural quantifier rules.

As can be gathered from the axioms, relational lattice theory is formulated in terms of pure logic. Therefore, if we take the universal closures of the axioms and call the collection Γ, the derivability relation of pure predicate logic can be used to express that A is a theorem of lattice theory: $\Gamma \vdash A$.

Lattice theory with operations has eight axioms, relational lattice theory instead twelve. We show that the former is 'at least as good' as the latter: Define the meet and join relations by:

$$M(a, b, c) \equiv a \wedge b = c \quad J(a, b, c) \equiv a \vee b = c$$

Axiom V is derived by:

$$\dfrac{\dfrac{\dfrac{\overline{a \wedge b \leqslant a \wedge b}^{\,Ref} \quad \overline{a \wedge b \leqslant a \wedge b}^{\,Ref}}{a \wedge b = a \wedge b}\,\&I}{\exists z\, a \wedge b = z}\,\exists I}{\forall x \forall y \exists z\, x \wedge y = z}\,\forall I, \forall I$$

By definition, we have proved $\forall x \forall y \exists z M(x, y, z)$. The rest of the relational axioms are derived similarly from the definition of meet and join.

Given a derivation in relational lattice theory, we can substitute the meet and join relations in it as in the above definition, then substitute the relational axioms by their derivations:

The meet and join relations can be defined, their axioms derived, and derivations replaced by ones in lattice theory with operations.

The language of relational lattice theory is not more expressive than that of lattice theory with operations, because the former can be emulated in the latter. In the other direction, to show the equivalence of the two axiomatizations, we proceed as follows: Let an atomic formula $t \leqslant s$ in lattice theory with operations be given. It is translated into relational lattice theory as follows: Let a, b be some ground terms in t, i.e., ones without lattice operations, such that also $a \wedge b$ is in t. Take a fresh term c and write down $M(a, b, c)$. If $a \wedge b$ was a component in some lattice term, say $(a \wedge b) \wedge d$, we now write $c \wedge d$ in its place and add $M(c, d, e)$ into our list, with e a fresh term. Proceeding in this way, we eventually find that t itself is some lattice term, say $e \wedge f$ and add $M(e, f, g)$ to our list. The same procedure with the term s gives us a list of relational atoms that finishes with, say, $M(e' \wedge f', h)$. We prove now by induction on the build-up of terms that if $t \leqslant s$ is provable in lattice theory with operations, then $g \leqslant h$ is provable from $M(a, b, c), \ldots, M(e, f, g), \ldots, M(e', f', h)$ in relational lattice theory. The procedure is completely general so that, combining the two ways, we have:

The equivalence theorem. *Derivations in relational lattice theory and lattice theory with operations can be translated to each other.*

The translations we have given exemplify the general method of translation between axiomatizations in terms of operations and relations.

12 | Elements of the proof theory of arithmetic

12.1 The Peano axioms

Giuseppe Peano published in 1889 an article with the title *Arithmetices principia, nova methodo exposita* (Latin for 'The principles of arithmetic, presented by a new method'). Peano took his axioms from Richard Dedekind's book of one year earlier, *Was sind und was sollen die Zahlen?* ('What are numbers and what are they for?'). Dedekind's aims were more general than Peano's, because he wanted to capture such notions as the infinity of a set. Peano's aim was to formalize arithmetic, and his work indeed gives a formal language that is the model of the notation standardly used today. Thus, Peano's notation was a big step ahead compared to that of Frege.

The famous five axioms of Peano read as follows, with the notation $a : \mathcal{N}$ to indicate that a is an element in $\mathcal{N}$:

Table 12.1 The axioms of Peano

I. $0 : \mathcal{N}$
II. $a : \mathcal{N} \supset a + 1 : \mathcal{N}$
III. $a : \mathcal{N} \,\&\, b : \mathcal{N} \supset (a = b \supset\subset a + 1 = b + 1)$
IV. $a : \mathcal{N} \supset \neg a + 1 = 0$
V. $A(0) \,\&\, \forall y(A(y) \supset A(y + 1)) \supset \forall x A(x)$

The notation is only slightly modernized: The sequence of natural numbers begins with 0 instead of 1 as in Peano, and the last axiom is written in first-order logic. Peano had followed Dedekind and his fifth axiom states that if for a class $\mathcal{K}$ we have that $0 : \mathcal{K}$ and $x : \mathcal{K} \supset x + 1 : \mathcal{K}$ for an arbitrary x, then $\mathcal{N}$ is contained in $\mathcal{K}$. To relate this principle to axiom V above, think of A as the class of all x that have the property A. To get rid of classes altogether in the formulation of the Peano axioms, we can take $\mathcal{N}$ to correspond to a predicate $N(x)$, 'to be a natural number', and write axioms I and II as $N(0)$ and $N(a) \supset N(a + 1)$, respectively.

A system of classical arithmetic, called **Peano arithmetic**, is obtained by adding the five Peano axioms to a system of classical predicate logic, with the rules of equality included. It is usual to include the possibility of defining functions by primitive recursion as explained below. Axioms I and II by which we can prove something to be a natural number are needed only for special purposes and can be left aside for now, along with the categorizations $a : \mathcal{N}, b : \mathcal{N}$ in axioms III and IV.

One direction of axiom III is the principle of replacement of equals in the successor function $x + 1$. The remaining direction, $a + 1 = b + 1 \supset a = b$ is an **axiom of infinity**. Dedekind had defined a class to be infinite if there is a one-to-one correspondence between the class and some proper subclass of it. From this perspective, axiom III guarantees that the image of $\mathcal{N}$ under the successor function s, written $s[\mathcal{N}]$ and abbreviated $\mathcal{N}^+$, has at least as many elements as $\mathcal{N}$: By axiom III, if two arguments a and b are distinct, the values $s(a)$ and $s(b)$ are distinct. The fourth axiom, $\neg\, s(a) = 0$, states that 0 is not the image of any number under s, so that $\mathcal{N}^+$ is a proper subclass of $\mathcal{N}$. For these reasons, axioms III and IV are called the two axioms of infinity.

The principle of arithmetic induction has been formulated in various ways. At school, one learns to prove the formula, reportedly used by Gauss in an intuitive way as a schoolboy, that gives the sum $1 + 2 + \ldots + n$ as $n(n + 1)/2$: First verify this for 1, then assume it for n and show that the same formula for $n + 1$ follows.

A variant formulation of induction is to show a property for 0, then to show that if for each $y \leq x$, the assumption $A(y)$ implies $A(x)$, then $\forall x A(x)$. Yet another version of the principle of induction is the method of 'infinite descent': To show $\forall x \neg A(x)$, show that $\neg A(0)$ and that $A(n + 1)$ implies $A(n)$. A restricted case of induction, sometimes called **special induction,** is the one in which a property is provable for 0 and for any successor sx. Thus, one infers $\forall x A(x)$ from $A(0)$ and $A(sx)$, with x arbitrary in the latter.

Philosophers of mathematics have tried to justify the principle of arithmetic induction, but the conclusion has been that it cannot be reduced to anything simpler. The intuitive motivation can be put as follows: Consider some property of natural numbers, such as $1 + 2 + \ldots + n = n(n + 1)/2$, and let this be denoted by $G(n)$. If $G(1)$ and $G(n) \supset G(n + 1)$ for any n, we can produce a proof of $G(m)$ for any specific given number m: Take $G(1)$ and instantiate $G(n) \supset G(n + 1)$ by the substitution $[1/n]$ to get $G(1) \supset G(2)$. Now $G(2)$ follows, so next instantiate $G(n) \supset G(n + 1)$ by $[2/n]$ to get $G(2) \supset G(3)$, and so on until you reach a proof of $G(m)$. Thus, the process terminates for any m, but the assumption that a proof of $G(m)$ is reached is just the principle of induction in disguise.

We should note that Peano's axiomatization of arithmetic is **schematic**: The principle of induction is meant as a scheme such that any of its instances, with A a concrete formula built up from atomic formulas by the connectives and quantifiers, is an axiom. It is not possible to tell in advance how these **induction formulas**, also called **inductive predicates**, should be chosen in proofs, neither is it possible in general to reduce the complexity of induction formulas. For example, if the induction formula is a conjunction and the task is to prove $A(sy)\ \&\ B(sy)$ from the assumption $A(y)\ \&\ B(y)$, the proof of just $A(sy)$ may require the use of $B(y)$. Therefore, the inductive step cannot be composed of two simpler steps, namely a proof of $A(sy)$ from $A(y)$ and a proof of $B(sy)$ from $B(y)$. An essential difference is met between quantified and quantifier-free induction formulas.

12.2 Heyting arithmetic

Peano arithmetic uses classical logic. If we change it into intuitionistic logic, we obtain **Heyting arithmetic**, so named after Arend Heyting who introduced it in connection with his axiomatization of intuitionistic logic in 1930. We shall formulate Heyting arithmetic, abbreviated **HA**, as an extension of the system of rules of intuitionistic natural deduction **NI**. We begin with the rules for equality:

Table 12.2 Rules for arithmetic equality

$$\frac{}{a=a}\,\textit{Ref} \qquad \frac{a=b}{b=a}\,\textit{Sym} \qquad \frac{a=b \quad b=c}{a=c}\,\textit{Tr}$$

Next, for each function introduced, a rule of replacement has to be added. For example, we have for the successor function the rule:

Table 12.3 Rule of replacement for the successor function

$$\frac{a=b}{s(a)=s(b)}\,\textit{s Rep}$$

A replacement rule is needed for all functions f, one that gives the conclusion $f(a)=f(b)$ from the premiss $a=b$. Therefore, one direction of Peano's third axiom is a principle of substitution of equals in the successor function. The schematic nature of the replacement rule leads to no problems, because the number of instances to be considered is the same as the number of functions that have been introduced. There is, however, another possible formulation in which the only functions are successor, sum, and product.

When functions are defined by **primitive recursion,** we add corresponding mathematical rules to the system. For sum and product, we have two replacement rules and two recursion rules for each operation:

Table 12.4 Replacement and recursion for sum and product

$$\frac{a = b}{a + c = b + c}{+Rep} \qquad \frac{b = c}{a + b = a + c}{+Rep}$$

$$\frac{}{a + 0 = a}{+0Rec} \qquad \frac{}{a + s(b) = s(a + b)}{+s\,Rec}$$

$$\frac{a = b}{a \times c = b \times c}{\times Rep} \qquad \frac{b = c}{a \times b = a \times c}{\times Rep}$$

$$\frac{}{a \times 0 = 0}{\times 0Rec} \qquad \frac{}{a \times s(b) = (a \times b) + b}{\times s\,Rec}$$

The addition of functions defined by recursion works always in the same way: We add for each function f replacement rules ($f Rep$) and zero-premiss **computation rules** for zero and successor ($f0Rec$ and *fs Rec*). Incidentally, the first or second replacement rule in sum and product can be left out if the commutativity of addition and multiplication is proved. Such proofs can be quite tricky and require the principle of induction. Therefore, in order to study the arithmetical rules with induction excluded, both replacement rules are added.

In setting up our system of intuitionistic arithmetic **HA**, we shall not consider the first two of Peano's axioms. As said, one half of axiom III is a principle of replacement, and the other half can be treated in at least two ways as we shall show. This leaves us with the fourth and fifth axioms, the former formulated as in:

Table 12.5 The rule of infinity

$$\frac{s(a) = 0}{\perp}{Inf}$$

The induction axiom is schematic, but in a way different from the replacement scheme, because there is an infinity of instances. We shall formulate it analogously to the E-rules of natural deduction, with an arbitrary consequence C and the abbreviation sy in place of $s(y)$:

Table 12.6 The induction rule

$$\dfrac{\begin{matrix}\Gamma\\ \vdots\\ A(0)\end{matrix}\qquad \begin{matrix}\overset{1}{A(y)},\Delta\\ \vdots\\ A(sy)\end{matrix}\qquad \begin{matrix}\overset{1}{A(t)},\Theta\\ \vdots\\ C\end{matrix}}{C}\,{\small \mathit{Ind},1}$$

Here A is any formula, and y is the eigenvariable of the rule, not free in Γ, Δ, Θ. In the derivation of the minor premiss C, the term t can be arbitrarily chosen. If t is a number instead of a variable, we have an instance of **numerical induction**. The rule of induction has no major premiss, but it has an arbitrary consequence and therefore behaves like a general elimination rule.

The standard rule of induction is obtained as a special case of the above rule, when $C \equiv A(t)$ and the degenerate derivation of the third premiss is left unwritten:

Table 12.7 The standard rule of induction

$$\dfrac{\begin{matrix}\Gamma\\ \vdots\\ A(0)\end{matrix}\qquad \begin{matrix}\overset{1}{A(y)},\Delta\\ \vdots\\ A(sy)\end{matrix}}{A(t)}\,{\small \mathit{Ind},1}$$

The restricted form of induction that was mentioned above, or 'special induction', has no assumption with an eigenvariable. Thus, special induction corresponds to the rule:

Table 12.8 The rule of special induction

$$\dfrac{\begin{matrix}\Gamma\\ \vdots\\ A(0)\end{matrix}\qquad \begin{matrix}\Delta\\ \vdots\\ A(sy)\end{matrix}\qquad \begin{matrix}\overset{1}{A(t)},\Theta\\ \vdots\\ C\end{matrix}}{C}\,{\small \mathit{Ind},1}$$

Special induction with the standard rule of Table 12.7 gives the conclusion $A(t)$ whenever A has been proved for zero and an arbitrary successor, $A(0)$ and $A(sy)$.

The system **HA** of Heyting arithmetic is defined as **NI** extended by rules for equality, replacement for successor, sum, and product, recursion for sum and product, and rules *Inf* and *Ind*. A natural deduction system of Peano arithmetic is obtained from **HA** by the addition of the principle of indirect proof for arbitrary formulas:

Definition 12.1. Heyting arithmetic.

$$\mathbf{HA} = \mathbf{NI} + \mathit{Ref} + \mathit{Sym} + \mathit{Tr} + \mathit{Rep} + \mathit{0Rec} + \mathit{sRec} + \mathit{Inf} + \mathit{Ind}.$$

Definition 12.2. Peano arithmetic.

$$\mathbf{PA} = \mathbf{HA} + \mathit{DN}.$$

Even if we have thus defined formal systems of proof, we shall often give informal proofs, instead of formal derivations by the rules of **HA** or **PA**. One reason is that such formal derivations become too broad to be printed. For better readability, we may use a dot in front of quantified equations, as in $\exists y.x = sy$.

One direction of Peano's third axiom is a principle of replacement, the other one is Dedekind's axiom of infinity, from $sa = sb$ to conclude $a = b$. We have:

Theorem 12.3. *If the terms a and b have no free variables, the rule by which one concludes $a = b$ from $sa = sb$ is admissible in* **HA**.

By this result, a proof of $sa = sb$ can be turned into a proof of $a = b$. There are at least two ways to the result, one of which is difficult, the other one easy: In the former, one shows by direct proof transformations that if $sa = sb$ has no free variables and is derivable, it is essentially concluded from $a = b$ (cf. Siders 2014). The easy way to the result is to have a system with arbitrary functions defined by primitive recursion and to define a **predecessor** function *prd* by the recursion equations:

$$prd(0) = 0 \qquad prd(sa) = a$$

If $sa = sb$, the principle of replacement for the predecessor function gives $prd(sa) = prd(sb)$ so $a = b$ follows by the above second recursion equation for *prd*.

An arithmetic axiom weaker than induction was proposed by Raphael Robinson in 1950:

Lemma 12.4. Robinson's axiom $\forall x(x = 0 \vee \exists y.x = sy)$ *is provable in* **HA**.

Proof. We prove $t = 0 \vee \exists y.t = sy$ for an arbitrary term t by induction. For the base case, we have $0 = 0 \vee \exists y.0 = sy$ by *Ref* and $\vee I_1$. For the inductive case, assume $x = 0 \vee \exists y.x = sy$ and show $sx = 0 \vee \exists y.sx = sy$.

If $x = 0$, then $sx = s0$ follows by *sRep*, so $\exists y.sx = sy$, consequently also $sx = 0 \vee \exists y.sx = sy$. If $\exists y.x = sy$, we apply $\exists E$ and assume $x = sz$ with z the eigenvariable of rule $\exists E$. Now, $sx = ssz$ by *sRep*, so $\exists y.sx = sy$ and $sx = 0 \vee \exists y.sx = sy$ follows. By $\vee E$, $sx = 0 \vee \exists y.sx = sy$, so by rule *Ind*, $t = 0 \vee \exists y.t = sy$, and $\forall I$ gives Robinson's axiom. QED.

The above proof suggests that **HA** is a natural collection of principles of proof. A comparison with a formal derivation shows how close to formalization the informal argument is. To keep the derivation manageable in breadth and readable in font size, we use the standard form of *Ind* in which the third minor derivation is degenerate (C equal to $A(t)$ in the schematic rule) and leave unwritten that minor derivation:

Table 12.9 A formal derivation of Robinson's axiom

$$
\dfrac{\dfrac{\dfrac{\dfrac{}{0 = 0}\,\scriptstyle Ref}{0 = 0 \vee \exists y.0 = sy}\,\scriptstyle \vee I_1 \qquad \dfrac{\dfrac{\overset{3}{x = 0 \vee \exists y.x = sy} \quad \dfrac{\dfrac{\overset{2}{x = 0}}{sx = s0}\,\scriptstyle sRep}{\exists y.sx = sy}\,\scriptstyle \exists I \quad \dfrac{\overset{2}{\exists y.x = sy} \quad \dfrac{\dfrac{\overset{1}{x = sz}}{sx = ssz}\,\scriptstyle sRep}{\exists y.sx = sy}\,\scriptstyle \exists I}{\exists y.sx = sy}\,\scriptstyle \exists E,1}{\exists y.sx = sy}\,\scriptstyle \vee E,2}{sx = 0 \vee \exists y.sx = sy}\,\scriptstyle \vee I_2}{v = 0 \vee \exists y.v = sy}\,\scriptstyle Ind,3}{\forall x(x = 0 \vee \exists y.x = sy)}\,\scriptstyle \forall I
$$

Robinson's axiom in place of the induction principle gives a weak system of arithmetic that nevertheless has a remarkable feature: It is finitely axiomatizable yet at the same time sufficiently strong for Gödel's **incompleteness theorem**, i.e., for the existence of a true arithmetic proposition that is not derivable within the formal system.

Theorem 12.5. *Excluded middle* $a = b \ \vee \neg\, a = b$ *for atoms is provable in* **HA**.

Proof. The proof is by induction on the first argument in the equation: We show first the base case $0 = y \ \vee \neg\, 0 = y$, for arbitrary y. By lemma 12.4, $y = 0$ or $\exists z.y = sz$. In the former case $0 = y \ \vee \neg\, 0 = y$ follows. In the latter case, *Inf* gives $\neg\, sz = 0$ so $\neg\, 0 = y$ follows by $sz = y$, therefore also $0 = y \ \vee \neg\, 0 = y$.

For the inductive case, assume $x = y \ \vee \neg\, x = y$ for an arbitrary y and show $sx = y \ \vee \neg\, sx = y$. By the assumption, following the easy method in the outline of proof of theorem 12.3, we get in particular the instance, $x = prd(y) \ \vee \neg\, x = prd(y)$. In the first case, if $x = prd(y)$, if $y = 0$, then $\neg\, sx = y$. If $x = prd(y)$, if $y = sz$ for some z, then by $s\,Rep$ and recursion for *prd*, $sx = y$. In the second case, if $\neg\, x = prd(y)$, then, assuming $sx = y$, we have $prd(sx) = prd(y)$ so $x = prd(y)$ which is impossible. Therefore $\neg\, sx = y$. All cases have now led to $sx = y \vee \neg\, sx = y$ so we conclude by induction that $a = b \vee \neg\, a = b$. QED.

Theorem 12.6. *Excluded middle $A \vee \neg A$ for quantifier-free formulas is provable in* **HA**.

Proof. Provability of excluded middle is shown by induction on the length of the formula. QED.

We shall now extend the normal form theorem for logical derivations to derivations in **HA**.

Theorem 12.7. *Given a derivation in* **HA**, *it can be converted into normal form.*

Proof. The proof is an extension of the normalization procedure of Section 13.2 below. We note first that the conclusions of all arithmetical rules except *Ind* are never major premisses in elimination rules, because they are atomic formulas. Therefore the only new case to consider in normalization is when such a major premiss has been derived by rule *Ind*. It is readily seen that the *E*-rule permutes up to the derivation of the third major premiss of rule *Ind*. QED.

Normal derivations begin with instances of the arithmetical rules, followed by logical rules and *Ind*. If in the rule of induction the arbitrary term t in the discharged assumption $A(t)$ is a number instead of a variable (case of 'numerical induction'), the instance of induction can be eliminated by repeated application of a conversion. The given derivation is:

$$\cfrac{\begin{matrix}\Gamma \\ \vdots \\ A(0)\end{matrix} \qquad \begin{matrix}\overset{1}{A(y)}, \Delta \\ \vdots \\ A(sy)\end{matrix} \qquad \begin{matrix}\overset{1}{A(t)}, \Theta \\ \vdots \\ C\end{matrix}}{C}\,{\scriptstyle Ind,1}$$

For simplicity, we assume the discharges simple. If $t = 0$, the conversion is into:

$$\begin{matrix}\Gamma \\ \vdots \\ A(0), \Theta \\ \vdots \\ C\end{matrix}$$

If $t = sn$, the conversion is into:

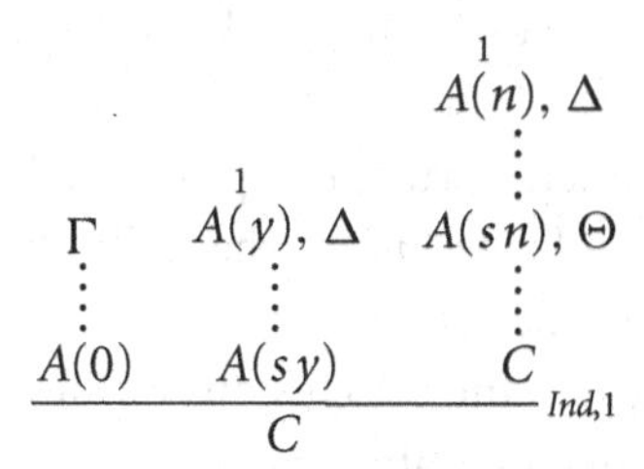

The derivation of $A(sn)$ from the assumption $A(n)$ is obtained by the substitution $[n/x]$ in the derivation of the second premiss. A repetition of the conversion will eventually produce a tower of composed derivations that begins with the derivation of $A(0)$ and continues to $A(1), \ldots, A(n), A(sn)$, with the induction eliminated. Note that open assumptions in Δ get multiplied, possibly zero times, in the conversion process.

Normal derivability in Heyting arithmetic has the same definition as in pure logic, namely that major premisses of elimination rules are assumptions. Contrary to the other arithmetical rules, rule *Ind* cannot be formulated so that it operates only on atomic formulas. As a consequence, there is no subformula property, and some authors write indeed for this reason that 'normalization fails in arithmetic'. Nevertheless, the normal form for arithmetical derivations is a useful property, as we shall now see.

12.3 The existence property

The Gödel–Gentzen translations of 1933, given in Section 10.3, were originally devised to reduce the consistency of classical arithmetic to intuitionistic arithmetic. The latter does not contain any of the 'dubitable' steps of indirect proofs with the quantifiers, but the question still remained whether intuitionistic arithmetic could be further justified as consistent. Reflecting on his incompleteness theorem, Gödel set out in a conference talk of 1933 a number of criteria for such a justification, including the existence property: If a formula $\exists x A(x)$ is a theorem in intuitionistic arithmetic, it should be possible to find an instance $A(n)$ on the basis of a proof of the theorem (cf. the lecture 'The present situation in the foundations of mathematics', printed in the third volume of Gödel's *Collected Works*).

The existence property of Heyting arithmetic was proved by Kleene in 1945 by what is known as the realizability method. We show as an application of the normal form of theorem 12.7 that the existence property can be proved by straightforward transformations of formal derivations of an existential theorem. The property follows with almost nothing to prove: Either the last rule in a normal derivation of $\exists x A(x)$ is $\exists I$ or the last rule can be dropped out.

We shall set the arithmetical rules except *Ind* aside now and consider the system of logical rules and rule *Ind* in what follows: As we have noted, the presence of arithmetical rules other than *Ind* does not affect the analysis of normal derivations because the conclusions of these arithmetical rules are never major premisses in elimination rules.

Theorem 12.8. Existence property for HA. *Given a derivation of $\exists x A(x)$ in* **HA** *with no open assumptions and no free variables in $\exists x A(x)$, a derivation of an instance $A(n)$ can be found.*

Proof. We may assume the derivation to be normal. E-rules would leave their major premisses as open assumptions, so the last rule is not an elimination. If it is $\exists I$, the premiss is some $A(t)$. If t is a number, a derivation of an instance is found by deleting the last rule, and if t is a term with a variable y, denoted $t(y)$, then the instance $A(t(n))$ is similarly derivable for any value of n.

By the above, we may assume that the last step in the derivation is *Ind*, with an induction formula B and the derivation:

$$\cfrac{\begin{matrix} & \overset{1}{B(z)} & \overset{1}{B(u)} \\ \vdots & \vdots & \vdots \\ B(0) & B(sz) & \exists x A(x) \end{matrix}}{\exists x A(x)}\ {\scriptstyle Ind,1}$$

Note that $B(0)$ is derivable without assumptions, so it is a theorem.

If u in the closed assumption $B(u)$ is a **closed term**, i.e., one without free variables, the resulting numerical induction can be removed as above. Therefore we can assume u to be a term $t(v)$ with a variable v. Consider the subderivation of the third minor premiss $\exists x A(x)$ from the assumption $B(u)$. Because v remains free in $B(u)$, it cannot be an eigenvariable of a rule instance in that subderivation. Therefore v can be substituted by 0 in the derivation of $\exists x A(x)$ from $B(u)$, with no assumptions left, and the instance of *Ind* that ends the derivation deleted. QED.

The condition that there be no free variables in $\exists x A(x)$ either forces the induction not to do any work, or else $\exists x A(x)$ has been concluded by rule $\exists I$.

We proved the disjunction property of intuitionistic logic in Section 3.4, and the existence property of intuitionistic predicate logic in Section 9.2. In

HA, the existence property leads to a definition of disjunction in terms of existence:

Corollary 12.9. *Disjunction is definable in* **HA** *by:*

$$A \vee B \equiv \exists x((x = 0 \supset A) \,\&\, (\neg\, x = 0 \supset B))$$

Proof. We show that the rules for $\vee$ follow from the definition. For $\vee I_1$, assume A. Then $0 = 0 \supset A$. Because $0 = 0$ is an instance of *Ref*, we have $\neg\, 0 = 0 \supset B$. Therefore $(0 = 0 \supset A) \,\&\, (\neg\, 0 = 0 \supset B)$ so $\exists I$ gives $\exists x((x = 0 \supset A) \,\&\, (\neg\, x = 0 \supset B))$, i.e., $A \vee B$. Rule $\vee I_2$ is derived similarly, with $(s0 = 0 \supset A) \,\&\, (\neg\, s0 = 0 \supset B)$. For $\vee E$, let the major premiss be $\exists x((x = 0 \supset A) \,\&\, (\neg\, x = 0 \supset B))$ and assume derivations of C from A and from B. By the existence property, $(n = 0 \supset A) \,\&\, (\neg\, n = 0 \supset B)$ for some n. By theorem 12.5, we have $n = 0 \vee \neg\, n = 0$. If $n = 0$, A follows, and therefore also C, and if $\neg\, n = 0$, B follows, and therefore also C, so that C follows. QED.

Corollary 12.10. Disjunction property for HA. *If $A \vee B$ is derivable in* **HA**, *with no free variables in A and B, then A or B is derivable.*

Proof. The result follows from the application of the existence property to $\exists x((x = 0 \supset A) \,\&\, (\neg\, x = 0 \supset B))$. QED.

The disjunction and existence properties fail for Peano arithmetic. One way to see the former is to consider a Gödelian unprovable proposition A. By the law of excluded middle, $A \vee \neg A$ is a theorem of **PA**, but by incompleteness, neither A nor $\neg A$ can be a theorem. On the other hand, if $A \vee \neg A$ were a theorem of **HA**, by the disjunction property either A or $\neg A$ would be, so that excluded middle for arbitrary formulas is not provable in **HA**. By the same, since $A \vee \neg A$ is provable in **HA** for quantifier-free A, no such A can be a Gödelian proposition.

The subformula property of normal derivations is lost when rule *Ind* is added to **NI**: All we can say is that formulas in a normal derivation are subformulas of open assumptions, of the conclusion, or of induction formulas.

The intuitive principle of arithmetic induction applies to any property $A(n)$ of natural numbers. There cannot be any effective enumeration of all possible arithmetic properties. What is known as the diagonal argument would produce new ones whenever an enumeration was suggested. One way to look at the incompleteness of arithmetic is that in a formal language there is a denumerable supply of arithmetic predicates, but these can never capture all possible properties of natural numbers. Thoralf Skolem came very close to trying to prove incompleteness on this basis, some years before Gödel. In a paper published in 1929, he wrote that 'it would be an interesting task to show that each collection of first-order formulas on the natural numbers remains valid when certain changes in the meaning of a "number" are made' (see Skolem 1929, p. 269).

12.4 A simple-minded consistency proof

We shall use the normal form for derivations in **HA** for a simple proof of the consistency of intuitionistic arithmetic, based on a notion of **falsifiability** of formulas. We show that if C is derivable from the assumptions Γ and if C is falsifiable, then some formula in Γ is falsifiable. Then, if no formula in Γ is falsifiable, neither is C.

Definition 12.11. Falsifiable formulas of HA. *If formula A has free variables, it is* **falsifiable** *when its universal closure is falsifiable. A collection of formulas Γ is* **falsifiable** *when Γ contains a falsifiable formula. For the rest, we have the inductive clauses:*

1. $\perp$ *is falsifiable.*
2. *A numerical equation $n = m$ is falsifiable if it is false.*
3. *$A \& B$ is falsifiable if A or B is falsifiable.*
4. *$A \vee B$ is falsifiable if A and B are falsifiable.*
5. *$A \supset B$ is falsifiable if A is not falsifiable whenever B is falsifiable.*
6. *$\forall x A(x)$ is falsifiable if $A(t)$ is falsifiable for some term t.*
7. *$\exists x A(x)$ is falsifiable if $A(t)$ is falsifiable for any term t.*

Clause 2 is justified by the decidability of numerical equality, theorem 12.3 above.

Theorem 12.12. Simple-minded consistency of HA. *If C is derivable from the assumptions Γ in* **HA** *and if C is falsifiable, then Γ is falsifiable.*

Proof. The proof is by induction on the height of a normal derivation of C. The base case is that C is an assumption, with $\Gamma \equiv C$, for which the claim is immediate. For the inductive case, assume the property for derivations up to height $\leqslant n$ and consider the last rule in derivations of height $\leqslant n + 1$:

1. The last rule is an arithmetical rule different from *Ind.* Zero-premiss rules do not have as conclusion falsifiable formulas, so the rule is *Sym, Tr, Rep,* or *Inf.* With *Sym,* if the conclusion is falsifiable, also the premiss is. With *Tr,* at least one premiss is falsifiable by $\neg a = c \supset \neg a = b \vee \neg b = c$, a formula derivable from *Tr* by the fact that classical logic can be applied, as in theorem 12.6. With *Rep*, the premiss is falsifiable, and with *Inf* likewise, through the false instance $s0 = 0$.

2. The last rule is $\&I$. The derivation, with $\Gamma \equiv \Gamma', \Gamma''$, is:

$$\dfrac{\begin{matrix}\Gamma' \\ \vdots \\ A\end{matrix} \quad \begin{matrix}\Gamma'' \\ \vdots \\ B\end{matrix}}{A \,\&\, B}\,{\scriptstyle \&I}$$

If $A \,\&\, B$ is falsifiable, one of A and B is, and by the inductive hypothesis, so is Γ' or Γ''.

3. The last rule is $\&E$. The derivation, with $\Gamma \equiv A \,\&\, B, \Gamma'$, is:

$$\dfrac{A \,\&\, B \qquad \begin{matrix}\overset{1}{A}, \overset{1}{B}, \Gamma' \\ \vdots \\ C\end{matrix}}{C}\,{\scriptstyle \&E,1}$$

By the inductive hypothesis, if C is falsifiable, then A or B or Γ' is falsifiable. In the two first cases $A \,\&\, B$ is falsifiable, so Γ is falsifiable in each case.

4. The last rule is $\vee I$ or $\vee E$. The proof is similar to **2** and **3**.

5. The last rule is $\supset I$. The derivation is:

$$\dfrac{\begin{matrix}\overset{1}{A}, \Gamma \\ \vdots \\ B\end{matrix}}{A \supset B}\,{\scriptstyle \supset I,1}$$

If $A \supset B$ is falsifiable, then A is not falsifiable whenever B is falsifiable. Therefore, by the inductive hypothesis, if B is falsifiable, also Γ is falsifiable.

6. The last rule is $\supset E$. The derivation, with $\Gamma \equiv A \supset B, \Gamma', \Gamma''$, is:

$$\dfrac{A \supset B \quad \begin{matrix}\Gamma' \\ \vdots \\ A\end{matrix} \quad \begin{matrix}\overset{1}{B}, \Gamma'' \\ \vdots \\ C\end{matrix}}{C}\,{\scriptstyle \supset E,1}$$

If C is falsifiable, B or Γ'' is falsifiable by the inductive hypothesis. In the former case, if A is falsifiable, then Γ' is, so if A is not falsifiable, then $A \supset B$ is by definition 12.11.

7. The last rule is $\bot E$. The derivation is

$$\cfrac{\begin{matrix}\Gamma\\ \vdots\\ \bot\end{matrix}}{C}\,{\scriptstyle \bot E}$$

Here Γ is falsifiable by the inductive hypothesis, because $\bot$ is.

8. The last rule is $\forall I$. The derivation is:

$$\cfrac{\begin{matrix}\Gamma\\ \vdots\\ A(y)\end{matrix}}{\forall x A(x)}\,{\scriptstyle \forall I}$$

If $\forall x A(x)$ is falsifiable, then $A(t)$ is falsifiable for some t. The substitution $[t/y]$ in the derivation of the premiss $A(y)$ gives a derivation of the falsifiable formula $A(t)$ from Γ so Γ is falsifiable by the inductive hypothesis.

9. The last rule is $\forall E$. The derivation, with $\Gamma \equiv \forall x A(x), \Gamma'$, is:

$$\cfrac{\forall x A(x) \qquad \begin{matrix}\overset{1}{A(t)}, \Gamma'\\ \vdots\\ C\end{matrix}}{C}\,{\scriptstyle \forall E,1}$$

If C is falsifiable, $A(t)$ or Γ' is falsifiable by the inductive hypothesis. With $A(t)$ falsifiable, also $\forall x A(x)$ is falsifiable.

10. The last rule is $\exists I$. The derivation is:

$$\cfrac{\begin{matrix}\Gamma\\ \vdots\\ A(t/x)\end{matrix}}{\exists x A(x)}\,{\scriptstyle \exists I}$$

If $\exists x A(x)$ is falsifiable, then $A(t)$ is falsifiable for any t, so Γ is falsifiable by the inductive hypothesis.

11. The last rule is $\exists E$. The derivation, with $\Gamma \equiv \exists x A(x), \Gamma'$, is:

$$\cfrac{\exists x A(x) \qquad \begin{matrix}\overset{1}{A(y)}, \Gamma'\\ \vdots\\ C\end{matrix}}{C}\,{\scriptstyle \exists E,1}$$

If C is falsifiable, $A(y)$ is falsifiable for an arbitrary y, or Γ' is falsifiable. In the former case, $\exists x A(x)$ is falsifiable.

12. The last rule is *Ind*. The derivation, with $\Gamma \equiv \Gamma', \Gamma', \Gamma'''$, is:

$$\cfrac{\begin{array}{c}\Gamma' \\ \vdots \\ A(0)\end{array} \quad \begin{array}{c}\overset{1}{A(y)}, \Gamma'' \\ \vdots \\ A(sy)\end{array} \quad \begin{array}{c}\overset{1}{A(t)}, \Gamma''' \\ \vdots \\ C\end{array}}{C}\,{\scriptstyle Ind, 1}$$

The case to consider is that $A(t)$ is falsifiable. By definition, $A(n)$ is then falsifiable for some number n. If $n = 0$, we conclude from the derivation of the first premiss $A(0)$ that Γ' is falsifiable. If $n = sm$, we do the substitution $[m/y]$ in the derivation of the second premiss, to get a derivation of $A(sm)$ from the assumptions $A(m), \Gamma''$. Here $A(sm)$ was assumed falsifiable, so the case to consider is when $A(m)$ is falsifiable. We reason as before that either $m = 0$ or $m = sk$, until we have for some number l the case $l = 0$ and conclude that Γ' is falsifiable. QED.

Corollary 12.13. **HA** *is consistent.*

Proof. An empty set of assumptions is not falsifiable. Therefore no theorem of **HA** is falsifiable. QED.

By Gödel's second incompleteness theorem, the proof of theorem 12.12 cannot be formalized in arithmetic. The proof itself is just fine, but the notion of falsifiability transcends elementary arithmetic: The problematic points of definition 12.11 are clauses 5, 6, and 7. In the case that there are free variables in $A \supset B$, falsifiability of $A \supset B$ connects the falsifiability of B to that of A: Those instances that falsify B should not falsify A. Moreover, this case leads to clause 6 by the falsifiability of open formulas (formulas with free variables). Both clauses 6 and 7 contain quantifiers ('some' and 'any') in the defining parts of the inductive clauses. These are the reasons why definition 12.11 goes beyond elementary arithmetic.

Theorem 12.12 and its proof could be formulated in a dual terminology of truth of formulas in **HA**:

Definition 12.14. **Truth in Heyting arithmetic.** *A formula in* **HA** *is* **true** *if it is not falsifiable.*

The falsifiability of formulas in arithmetic is not arithmetically definable, so neither is truth. If Γ consists of true assumptions and C is derivable from Γ, then C is true. A result to this effect is announced in Kleene's book *Introduction to Metamathematics* (theorem 61, p. 500), with the indication that a proof can be constructed in analogy to a proof of the consistency of the predicate calculus.

PART IV

Complementary topics

13 | Normalization and cut elimination

13.1 Proofs by structural induction

(a) **Inductive generation.** A class of objects is said to be **inductively generated** if, given some finite collection of objects and operations on objects, each object in the class is obtained from the finite collection after some bounded number of operations.

The importance of inductive generation is that it gives us a systematic means of proving properties of infinite collections of objects: The principle is to first prove that each object in the given finite collection has the property, then to prove that each way of generating new objects by the operations maintains the property.

The above abstract description is best illustrated by examples. The first and best-known inductive class is that of the natural numbers. The given finite collection is just the number 0, and there is a single operation, what is called the **successor** of a number n, the application of which is written $s(n)$. The standard notation is 1 for $s(0)$, 2 for $s(s(0))$, etc.

The principle of proof that goes with an inductive generation of the natural numbers is called **arithmetic induction**, or 'complete induction' in the older literature: Given a property of natural numbers, written $A(n)$ for n, prove that 0 (or sometimes 1) has the property and that if n has the property, also $s(n)$ has. In other words, prove $A(0)$ and $A(n) \supset A(s(n))$. If n is arbitrary, the conclusion is that every natural number has the property.

Many inductive proofs are so simple that they go practically unrecognized, as with the property expressed by: *One out of three consecutive natural numbers is divisible by 3.* We see that 3 is divisible by 3, 6 is divisible by 3, 9 is divisible by 3, and so on, an insight used in example 1.2(c). Yet, what we see in our mind's eye would not count for a mathematical proof. We want to prove by arithmetical means that for any n, one of n, $n+1$, $n+2$ is divisible by 3. For the proof by induction, let first $n = 1$, and we have that one of 1, $1+1$, $1+2$ is divisible by 3. Next assume the property for n, i.e., assume that one of n, $n+1$, $n+2$ is divisible by 3, and prove the property for $n+1$, i.e., prove that one of $n+1$, $n+2$, $n+3$ is divisible by 3. The assumption gives three cases, and the divisibility by 3 of $n+1$ and $n+2$

leads to the claim. The remaining case is that n is divisible by 3, by which also $n+3$ is divisible by 3: There is by assumption a k such that $n = 3k$, by which $n+3 = 3k+3 = 3(k+1)$.

In the inductive proof, we have the **base case** with 0 (or, as in the above, with 1), and the **inductive case** with the step from n to $s(n)$. In the latter, n has to be arbitrary, i.e., an **eigenvariable** in the sense of predicate logic (cf. Section 8.2): Nothing must be assumed about n except that it is a natural number.

We can write the principle of induction as a rule that can be added to the system of natural deduction for predicate logic:

$$\dfrac{\begin{matrix}\Gamma \\ \vdots \\ A(0)\end{matrix} \qquad \begin{matrix}\overset{1}{A(n)}, \Delta \\ \vdots \\ A(s(n))\end{matrix}}{\forall x A(x)}\,{\scriptstyle Ind,1}$$

The temporary assumption $A(n)$ is closed at the inference. The number n in it has to be arbitrary, i.e., no assumption about n must be contained in Γ, Δ. Special induction is that special case of rule *Ind* in which the discharge of $A(n)$ is vacuous.

Given an inductive class such as the natural numbers, one can define new operations, such as the sum of two numbers:

1. $n+0 = n$
2. $n+s(m) = s(n+m)$

In this **recursive definition** of the sum, it is first defined what the addition of 0 gives, then what the addition of a successor $s(m)$ gives. The latter is reduced to the addition of m that is smaller than $s(m)$. The cases for m are two:

1. If $m = 0$, we have $s(n+m) = s(n+0) = s(n)$.
2. If $m = s(k)$ for some k, we have $s(n+m) = s(n+s(k)) = s(s(n+k))$.

With k, the cases are again two, until the second term in the sum is 0 and vanishes as in case 1.

We can now prove inductively properties of the defined operation, such as $n+m = m+n$, for every n, m. Such proofs can be rather involved; for example, sum is an operation that takes two arguments, and repeated sums are written with parentheses. A proof of the commutativity of sum requires first a proof of the associative law $(n+m)+k = n+(m+k)$.

The product of two numbers is defined by:

1. $n \times 0 = 0$
2. $n \times s(m) = n \times m + n$

We can now prove properties of natural numbers, such as the formula $2 \times (1 + 2 + \ldots + n) = n \times (n + 1)$ that Gauss is said to have discovered as a child when given the task to do the sum $1 + 2 + \ldots + 100$.

(b) Structural induction. The natural numbers are the historically first inductive class. In logic, the inductive classes of greatest interest are those of formulas and derivations. The formulas of propositional logic were defined in Section 1.6 as follows: First the class of simple formulas was given, then the inductive clauses by which compound formulas can be constructed. Thus, we have:

Table 13.1 Inductive generation of formulas

1. $P, Q, R, \ldots$ and $\bot$ are **formulas.**
2. If $A, B, C, \ldots$ are **formulas,** also $(A) \,\&\, (B)$, $(A) \vee (B)$, and $(A) \supset (B)$ are **formulas.**

In each concrete situation, the number of simple formulas, i.e., those under clause 1, is bounded, but we cannot put any fixed bound on their number. (In the end, the natural numbers are required.)

The principle of proof based on the inductive generation of formulas from the simple ones is called **induction on the length** of a formula, where length $l(A)$ of a formula A is the number of connectives in it:

$$l(P) = 0$$
$$l(\bot) = 1$$
$$l((A) \,\&\, (B)) = l((A) \vee (B)) = l((A) \supset (B)) = l(A) + l(B) + 1$$

For an unimaginative example of such a proof, consider *All compound formulas have an even number of parentheses.* The base case of structural induction is a simplest compound formula. It has four parentheses, as is seen from the three operations for forming compound formulas. Next, for the inductive step, assume both of A and B to have an even number of parentheses. Each of the three constructions adds four parentheses, by which the conclusion follows.

Induction on the length of a formula was used in the proof of the admissibility of the rule of contraction in Section 4.5.

A formal system of proof is nothing but an inductive definition of a class of derivable formulas or, more generally, of formulas derivable from given assumptions. As an example, consider intuitionistic sequent calculus for propositional logic:

Table 13.2 Inductive generation of derivations

1. The sequents $A, \Gamma \vdash A$, and $\bot, \Gamma \vdash C$ are **derivations** of $A, \Gamma \vdash A$, and $\bot, \Gamma \vdash C$.
2. Given **derivations** of the sequents found as premisses of the rules of Table 4.1, application of the rules produces **derivations** of the sequents found as conclusions of these rules.

Derivations start with initial sequents, followed by applications of the logical rules. Thus, the formal definition of derivations goes in a direction opposite to that of finding derivations. The latter proceeds in a root-first direction. Structural induction on the **height** of a derivation proceeds in the direction in which derivations are defined, i.e., it starts from the initial sequents that form the leaves of derivation trees. Height is defined as the greatest number of consecutive steps in a derivation tree, i.e., as the length of its longest branch. Induction on the height of a derivation was used first in the proof of the invertibility of rules in sequent calculus, in Section 4.2.

Looking at the proof of admissibility of contraction in Section 4.5, we notice that it is a 'double induction', with two **inductive parameters**: The main or **principal** induction is on the length of the contraction formula, the secondary one, the **subinduction**, on the height of derivation. Derivations were transformed in the proof so that either contraction was applied on a shorter formula, with no control on the height of derivation, or height was reduced but the length of the contraction formula kept intact. We can depict the parameter as a pair (n, m). The main induction shows that n can be reduced but says nothing of what happens to m. The subinduction shows that m can be reduced with n left intact. Altogether, we arrive at the case of a simple formula $(n = 0)$ and an initial sequent $(m = 0)$.

13.2 A proof of normalization

We give a proof of normalization for a system of intuitionistic natural deduction. It will be useful to write derivations in **sequent calculus style**, as defined by the following translation of the system of rules of Sections 3.4 (rules for connectives) and 9.1 (quantifier rules).

Root-first translation of natural deduction into sequent calculus style:

$$\dfrac{\begin{array}{c}\Gamma\\ \vdots\\ A\end{array}\quad\begin{array}{c}\Delta\\ \vdots\\ B\end{array}}{A\,\&\,B}\,\&I \qquad\qquad \dfrac{\begin{array}{c}\vdots\\ \Gamma\vdash A\end{array}\quad\begin{array}{c}\vdots\\ \Delta\vdash B\end{array}}{\Gamma,\Delta\vdash A\,\&\,B}\,\&I$$

$\vee I$ is similar, and $\supset I$ is:

$$\dfrac{\begin{array}{c}\overset{1}{A^n},\Gamma\\ \vdots\\ B\end{array}}{A\supset B}\supset I \qquad\qquad \dfrac{\begin{array}{c}\vdots\\ A^n,\Gamma\vdash B\end{array}}{\Gamma\vdash A\supset B}\supset I$$

Rule $\vee E$ is translated as:

$$\dfrac{A\vee B\quad\begin{array}{c}\overset{1}{A^n}\\ \vdots\\ C\end{array}\quad\begin{array}{c}\overset{1}{B^m}\\ \vdots\\ C\end{array}}{C}\vee E,1 \qquad\qquad \dfrac{\Gamma\vdash A\vee B\quad A^n,\Delta\vdash C\quad B^m,\Theta\vdash C}{\Gamma,\Delta,\Theta\vdash C}\vee E$$

The translation continues from the premisses until assumptions are reached.

The logical rules of the calculus **NLI** are obtained by translating the rest of the logical rules in sequent notation. The nomenclature **NLI** was used in some early manuscripts of Gentzen to denote a 'natural-logistic intuitionistic calculus'.

Table 13.3 Natural deduction in sequent style

$\dfrac{\Gamma\vdash A\,\&\,B\quad A^n,B^m,\Delta\vdash C}{\Gamma,\Delta\vdash C}\&E$	$\dfrac{\Gamma\vdash A\quad\Delta\vdash B}{\Gamma,\Delta\vdash A\,\&\,B}\&I$
$\dfrac{\Gamma\vdash A\vee B\quad A^n,\Delta\vdash C\quad B^m,\Theta\vdash C}{\Gamma,\Delta,\Theta\vdash C}\vee E$	$\dfrac{\Gamma\vdash A}{\Gamma\vdash A\vee B}\vee I_1\qquad\dfrac{\Gamma\vdash B}{\Gamma\vdash A\vee B}\vee I_2$
$\dfrac{\Gamma\vdash A\supset B\quad\Delta\vdash A\quad B^m,\Theta\vdash C}{\Gamma,\Delta,\Theta\vdash C}\supset E$	$\dfrac{A^n,\Gamma\vdash B}{\Gamma\vdash A\supset B}\supset I$
$\dfrac{\Gamma\vdash\forall xA(x)\quad A(t)^n,\Delta\vdash C}{\Gamma,\Delta\vdash C}\forall E$	$\dfrac{\Gamma\vdash A(y)}{\Gamma\vdash\forall xA(x)}\forall I$
$\dfrac{\Gamma\vdash\exists xA(x)\quad A(y)^n,\Delta\vdash C}{\Gamma,\Delta\vdash C}\exists E$	$\dfrac{\Gamma\vdash A(t)}{\Gamma\vdash\exists xA(x)}\exists I$

The calculus is completed by adding initial sequents of the form $A\vdash A$, with A an arbitrary formula, and the zero-premiss rule $\bot E$ by which $\bot\vdash C$ can begin a derivation branch.

We say that the closing of assumption formulas in E-rules and in rule $\supset I$ is **vacuous** if $n=0$ or $m=0$. Similarly, the closing of an assumption is **multiple** if $n > 1$ or $m > 1$. With $n = 1$ or $m = 1$, the closing of an assumption is **simple**. Vacuous and multiple closing of assumptions is seen in standard natural deduction in:

$$\dfrac{\begin{array}{c}\Gamma\\ \vdots\\ B\end{array}}{A \supset B}{\scriptstyle \supset I} \qquad \dfrac{\begin{array}{c}\overset{1}{A}, \overset{1}{A}, \Gamma\\ \vdots\\ B\end{array}}{A \supset B}{\scriptstyle \supset I,1}$$

The former case corresponds to the situation in sequent calculus in which a formula active in a logical rule stems from a step of weakening, the latter to a situation in which it stems from a step of contraction (cf. *Structural Proof Theory*, p. 175).

The sequent calculus for root-first proof search of Chapter 4, Table 4.1, is more restricted than **NLI**. If in the latter the major premisses of E-rules are assumptions, the sequent notation gives $A \& B \vdash A \& B$ for the major premiss, and similarly for the other E-rules. When these are left unwritten and the multiplicities in closed assumptions are simple, the sequent calculus for root-first proof search is obtained, save for the shared contexts in two-premiss rules (cf. also Section 13.4).

In a **permutative conversion**, the height of derivation of a major premiss derived by $\vee E$ or $\exists E$ is diminished. The effect of the general rules is that such conversions work for all derived major premisses of elimination rules, by which:

Definition 13.1. *A derivation in natural deduction with general elimination rules is* **normal** *if all major premisses of E-rules are assumptions.*

The **composition** of two derivations is an essential step in the normalization of derivations. It can now be written quite generally in the form:

$$\frac{\Gamma \vdash D \quad D, \Delta \vdash C}{\Gamma, \Delta \vdash C}{\scriptstyle Comp}$$

Iterated compositions appear as so many successive instances of rule *Comp*.

Lemma 13.2. Closure of derivations with respect to composition. *If given derivations of the sequents $\Gamma \vdash D$ and $D, \Delta \vdash C$ in* **NLI** *are composed by rule* Comp *to conclude the sequent $\Gamma, \Delta \vdash C$, the instance of* Comp *can be eliminated.*

Proof. We show by induction on the height of derivation of the right premiss of *Comp* that it can be eliminated.

1. Base case. The second premiss of *Comp* is an initial sequent, as in:

$$\frac{\Gamma \vdash D \quad D \vdash D}{\Gamma \vdash D}\,{\scriptstyle Comp}$$

The conclusion of *Comp* is identical to its first premiss, so that *Comp* can be deleted.

If the second premiss is of the form $\bot \vdash D$, the first premiss is $\Gamma \vdash \bot$. It has not been derived by a right rule, so that *Comp* can be permuted up in the first premiss. In the end, a topsequent $\Gamma' \vdash \bot$ is found as the left premiss of *Comp*, by which $\bot$ is in Γ', so that the conclusion of *Comp* is an initial sequent.

2. Inductive case with the second premiss of *Comp* derived by an I-rule. There are two subcases, a one-premiss rule and a two-premiss rule. In the former case, *Comp* is permuted up to the premiss, with a lesser height of derivation as a result. In the latter case, we use the notation (D) to indicate a possible occurrence of D in a premiss:

$$\frac{\Gamma \vdash D \qquad \dfrac{(D), \Delta' \vdash C' \quad (D), \Delta'' \vdash C''}{D, \Delta', \Delta'' \vdash C}\,{\scriptstyle Rule}}{\Gamma, \Delta', \Delta'' \vdash C}\,{\scriptstyle Comp}$$

Rule *Comp* is permuted to any premiss that has an occurrence of D, say the first one, with the result:

$$\frac{\dfrac{\Gamma \vdash D \quad D, \Delta' \vdash C'}{D, \Delta' \vdash C'}\,{\scriptstyle Comp} \qquad \Delta'' \vdash C''}{\Gamma, \Delta', \Delta'' \vdash C}\,{\scriptstyle Rule}$$

3. Inductive case with the second premiss of *Comp* derived by an E-rule, as in:

$$\frac{\Gamma \vdash D \qquad \dfrac{(D), \Delta \vdash A \& B \quad (D), A^n, B^m, \Theta \vdash C}{D, \Delta, \Theta \vdash C}\,{\scriptstyle \&E}}{\Gamma, \Delta, \Theta \vdash C}\,{\scriptstyle Comp}$$

As in case 2, *Comp* is permuted up, to whichever premiss has an occurrence of the composition formula D, with a lesser height of derivation as a result. QED.

In the case of a multiple discharge, a detour conversion will lead to several compositions, with a multiplication of the contexts as in the example:

$$\dfrac{\dfrac{\Gamma \vdash A \quad \Delta \vdash B}{\Gamma, \Delta \vdash A \& B}\,\&I \qquad A, A, B, \Theta \vdash C}{\Gamma, \Delta, \Theta \vdash C}\,\&E$$

The conversion is into:

$$\dfrac{\Delta \vdash B \qquad \dfrac{\Gamma \vdash A \qquad \dfrac{\Gamma \vdash A \quad A, A, B, \Theta \vdash C}{A, B, \Gamma, \Theta \vdash C}\,Comp}{B, \Gamma, \Gamma, \Theta \vdash C}\,Comp}{\Gamma, \Gamma, \Delta, \Theta \vdash C}\,Comp$$

Such multiplication does not affect the normalization process. Note well that normalization depends on the admissibility of composition, which latter has to be proved **before** normalization.

In normalization, derived *MP*'s of E-rules are converted step by step into assumptions. There are two situations, depending on whether the *MP* was derived by an E-rule or an I-rule.

Definition 13.3. Normalizability. *A derivation in* **NLI** *is* **normalizable** *if there is a sequence of conversions that transform it into normal form.*

The idea of our proof of the normalization theorem is to show by induction on the last rule applied in a derivation that logical rules maintain normalizability.

The cut elimination theorem, to be presented in Section 13.4, is often called *Gentzen's Hauptsatz*, main theorem. He used the word *Hilfssatz*, auxiliary theorem or lemma, for an analogous result by which composition of derivable sequents maintains the reducibility of sequents, a property defined in his original proof of the consistency of arithmetic in 1935 (cf. von Plato 2014 for details). Here we have as a crucial element of the normalization theorem an analogous *Hilfssatz* by which normalizability is maintained under composition.

Theorem 13.4. Normalizability for intuitionistic natural deduction. *Derivations in* **NLI** *convert to normal form.*

Proof. Consider the last rule applied. The base case is an assumption that is a normal derivation. In the inductive case, if an I-rule is applied to premisses the derivations of which are normalizable, the result is a normalizable derivation. The same holds if a normal instance of an E-rule is applied. The remaining case it that a non-normal instance of an E-rule is applied. The *MP* of the rule is then derived either by another E-rule or an I-rule, so we have two main cases with subcases according to the specific rule in each. Derivations are so transformed that normalizability of the last rule instance can be concluded either because the rule instance resolves into possible non-normalities with shorter conversion formulas, or because the height of derivation of its major premiss is diminished.

1. *E-rules:* Let the rule be $\&E$ followed by another instance of $\&E$, as in:

$$\cfrac{\cfrac{\Gamma \vdash A \& B \quad A^n, B^m, \Delta \vdash C \& D}{\Gamma, \Delta \vdash C \& D}{\scriptstyle \&E} \quad C^k, D^l, \Theta \vdash E}{\Gamma, \Delta, \Theta \vdash C \& D}{\scriptstyle \&E}$$

By the inductive hypothesis, the derivations of the premisses of the last rule are normalizable. The second instance of $\&E$ is permuted above the first:

$$\cfrac{\Gamma \vdash A \& B \quad \cfrac{A^n, B^m, \Delta \vdash C \& D \quad C^k, D^l, \Theta \vdash E}{A^n, B^m, \Delta, \Theta \vdash E}{\scriptstyle \&E}}{\Gamma, \Delta, \Theta \vdash E}{\scriptstyle \&E}$$

The height of derivation of the *MP* of the last rule instance in the upper derivation has diminished by 1, so the subderivation down to that rule instance is normalizable. The height of the *MP* of the other rule instance has remained intact and is therefore normalizable.

All other cases of permutative convertibility go through in the same way.

2. *I-rules:* The second situation of convertibility is that the *MP* has been derived by an I-rule, as in:

$$\cfrac{\cfrac{\Gamma \vdash A \quad \Delta \vdash B}{\Gamma, \Delta \vdash A \& B}{\scriptstyle \&I} \quad A^n, B^m, \Theta \vdash C}{\Gamma, \Delta, \Theta \vdash C}{\scriptstyle \&E}$$

Let us assume for the time being that $n = m = 1$. The detour conversion is given by:

$$\cfrac{\Delta \vdash B \quad \cfrac{\Gamma \vdash A \quad A, B, \Theta \vdash C}{B, \Gamma, \Theta \vdash C}{\scriptstyle Comp}}{\Gamma, \Delta, \Theta \vdash C}{\scriptstyle Comp}$$

The result is not a derivation in **NLI**. We proved in Lemma 13.2 that *Comp* is eliminable. The next step is to show that *Comp* maintains normalizability. This will be done in the *Hilfssatz* to be proved separately. By the *Hilfssatz*, the conclusion of the upper *Comp* is normalizable, and again by the *Hilfssatz*, also the conclusion of the lower *Comp*. If $n > 1$ or $m > 1$, *Comp* is applied repeatedly, the admissibility of an uppermost *Comp* giving the admissibility of the following ones.

The other I-rules go through in the same way. Just to check, let us go through rule $\supset I$. The detour convertibility is:

$$\dfrac{\dfrac{A^n, \Gamma \vdash B}{\Gamma \vdash A \supset B}{\scriptstyle \supset I} \quad \Delta \vdash A \quad B^m, \Theta \vdash C}{\Gamma, \Delta, \Theta \vdash C}{\scriptstyle \supset E}$$

In the conversion, multiple discharge of assumptions is again resolved into iterated compositions, so we may assume $n = m = 1$ and have the conversion:

$$\dfrac{\dfrac{\Delta \vdash A \quad A, \Gamma \vdash B}{\Gamma, \Delta \vdash B}{\scriptstyle Comp} \qquad B, \Theta \vdash C}{\Gamma, \Delta, \Theta \vdash C}{\scriptstyle Comp}$$

By the *Hilfssatz*, the derivation is normalizable. QED.

It remains to give a proof of the *Hilfssatz:*

Hilfssatz 13.5. Closure of normalizability under composition. *If the premisses of rule* Comp *are normalizable, also the conclusion is.*

Proof. The proof is by induction on the length of the composition formula D with a subinduction on the sum of the heights of derivation of the two premisses.

1. $D \equiv P$. With an atomic formula P, we have:

$$\dfrac{\Gamma \vdash P \quad P, \Delta \vdash C}{\Gamma, \Delta \vdash C}{\scriptstyle Comp}$$

P is never principal in the right premiss, so that *Comp* can be permuted up with a lesser sum of heights of derivation as a result. There are two cases, a one-premiss rule and a two-premiss rule. For the latter, we use again the notation (P) to indicate a possible occurrence of P in a premiss:

$$\dfrac{\Gamma \vdash P \qquad \dfrac{(P), \Delta' \vdash C' \quad (P), \Delta'' \vdash C''}{P, \Delta', \Delta'' \vdash C}{\scriptstyle Rule}}{\Gamma, \Delta', \Delta'' \vdash C}{\scriptstyle Comp}$$

Rule *Comp* is permuted to the premiss that has an occurrence of P, say the first one, with the result:

$$\dfrac{\dfrac{\Gamma \vdash P \quad P, \Delta' \vdash C'}{P, \Delta' \vdash C'}{\scriptstyle Comp} \qquad \Delta'' \vdash C''}{\Gamma, \Delta', \Delta'' \vdash C}{\scriptstyle Rule}$$

In the end, the second premiss of *Comp* is an initial sequent, as in:

$$\frac{\Gamma \vdash P \quad P \vdash P}{\Gamma \vdash P}\,\textit{Comp}$$

The conclusion of *Comp* is identical to its first premiss, so that *Comp* can be deleted.

2. $D \equiv \bot$. Because $\bot$ is never principal in the left premiss, *Comp* is permuted up as in the proof of admissibility of composition.

3. $D \equiv A \& B$. If $A \& B$ is not principal in the right premiss, *Comp* can be permuted as in 1.

If $A \& B$ is principal, there has to be a normal rule instance in the right premiss, as in:

$$\frac{\Gamma \vdash A \& B \qquad \dfrac{A \& B \vdash A \& B \quad A^n, B^m, \Delta \vdash C}{A \& B, \Delta \vdash C}\,\&E}{\Gamma, \Delta \vdash C}\,\textit{Comp}$$

Comp is permuted up to the first premiss:

$$\frac{\dfrac{\Gamma \vdash A \& B \quad A \& B \vdash A \& B}{\Gamma \vdash A \& B}\,\textit{Comp} \qquad A^n, B^m, \Delta \vdash C}{\Gamma, \Delta \vdash C}\,\&E$$

Comp is now deleted and a generally non-normal instance of rule $\&E$ created. If the major premiss is concluded by an E-rule, a permutative conversion is done and no instance of *Comp* created. If the last rule is $\&I$, a detour convertibility with the conversion formula $A \& B$ is created. A detour conversion will lead to new instances of *Comp*, but on strictly shorter formulas.

The other cases of composition formulas are treated in a similar way. QED.

The proof is extended without problems to the calculus of natural deduction for classical logic.

13.3 The Curry–Howard correspondence

At the time when intuitionistic logic was in its infancy, around 1930, Heyting and Kolmogorov suggested an explanation of the principles of intuitionistic logic in terms of the notion of proof. Also Gentzen makes in his doctoral thesis the suggestion that the introduction rules give the meanings of the various forms of propositions in terms of provability. These rules make more precise Heyting's earlier discussion. In Kolmogorov (1932), it is suggested that intuitionistic logic is a 'logic of problem solving': The atomic formulas express the primitive problems that have no logical structure. $A \& B$ expresses a problem that is solved by solving A and B separately, $A \vee B$ is solved by solving at least one of A and B, and $A \supset B$ is solved by reducing the solution of B to one of A. Falsity $\bot$ expresses an impossible problem that has no solution. In Kolmogorov's interpretation, the notion of a problem comes before the notion of a theorem: A theorem can be considered that special case of a problem in which the task is to find a proof.

(a) Typed lambda-calculus. In 1969, William Howard made precise some of the ideas behind the proof interpretation of intuitionistic logic, in his article 'The formulae-as-types notion of construction'. The paper circulated as a manuscript and was finally published as Howard (1980). It established what came to be called the 'Curry–Howard isomorphism' or 'Curry–Howard correspondence'. Curry's role was that he suggested the idea for implication in his book (1958). The basic idea is that a formula corresponds to the set of its proofs. More precisely, to each formula A there is the set of proofs of A in the sense of formal derivation. The notation $a : A$ stands for 'a is a proof of A'. In terms of sets, the reading is 'a is an element of the set A'. An introduction rule shows how to construct a proof from proofs of components: If $a : A$ and $b : B$, then $pair(a, b) : A \& B$. The operation of forming $pair(a, b)$ is the construction that gives a proof of the conjunction $A \& B$. If $a : A$, then $i(a) : A \vee B$, if $b : B$, then $j(b) : A \vee B$. The two operations indicated by i and j carry the information from which component of the disjunction the proof is constructed, proof of A or proof of B. Implication is more difficult: Assume an arbitrary proof or a 'witness' w of A, so symbolically $w : A$. If you succeed in constructing from w a proof $b(w)$ of B, then the proof of $A \supset B$ is written as $\lambda w\, b(w)$. This is the **lambda-abstract** of the expression $b(w)$ that depends on the variable w, as invented by Alonzo Church (1932). Set-theoretically, $A \& B$ is the Cartesian product of the sets A, B, and $A \vee B$ their 'disjoint' union, and $A \supset B$ the set of functions from A to B.

As to the elimination rules, they show how to pass from an arbitrary proof of a formula to proofs of its components: If $w : A \& B$, then $p(w) : A$ and $q(w) : B$ are the projection constructions that do this. For implication, using a suggestive symbol for a member of $A \supset B$, if $f : A \supset B$ and $w : A$, then $f(w) : B$. In terms of sets, a proof f of $A \supset B$ is a function f that transforms any proof w of A into some proof $f(w)$ of B. Thus, rule *modus ponens* is the same as the **application** of a function. Implication introduction, then, is **functional abstraction** as invented by Church.

The notion of proof has been interpreted in a very specific way here, namely so that proofs are represented by what are often called the **proof terms** of a formal calculus of terms. A proof term should indicate what rule produced it, and therefore we write $app(f, a) : B$ whenever rule $\supset E$ was used, instead of simply $f(a) : B$, and similarly for all the other terms.

The rules of natural deduction become the rules of **typed lambda-calculus** under the Curry–Howard correspondence:

Table 13.4 Typed natural deduction rules

$$\frac{a : A \quad b : B}{pair(a, b) : A \& B}\,\&I \qquad \frac{c : A \& B}{p(c) : A}\,\&E_1 \qquad \frac{c : A \& B}{q(c) : B}\,\&E_2$$

$$\frac{\begin{array}{c} \overset{1}{w : A} \\ \vdots \\ b(w) : B \end{array}}{\lambda w\, b(w) : A \supset B}\,{\supset}I,1 \qquad \frac{f : A \supset B \quad a : A}{app(f, a) : B}\,{\supset}E$$

$$\frac{a : A}{i(a) : A \vee B}\,\vee I_1 \qquad \frac{b : B}{j(b) : A \vee B}\,\vee I_2 \qquad \frac{c : A \vee B \quad \begin{array}{c} \overset{1}{w : A} \\ \vdots \\ a(w) : C \end{array} \quad \begin{array}{c} \overset{1}{v : B} \\ \vdots \\ b(v) : C \end{array}}{d(c, \lambda w\, a(w), \lambda v\, b(v)) : C}\,\vee E,1$$

In rule $\supset I$, an arbitrary proof w is assumed for A, then a proof $b(w)$ constructed for B that depends on the witness w. The notation $\lambda w\, b(w)$ stands for the functional abstract that gives the proof of $A \supset B$. Rule $\vee E$ has similarly two assumptions, a proof w of A and a proof v of B, and two functions that correspond to the derivations of C from each assumption. (We have simplified the presentation a bit here by using λ in rule $\vee E$, cf. *Structural Proof Theory*, appendix B.)

In Table 13.4, the **terms** at the left of the colon that correspond to introduction rules are often called **constructors** and those that correspond to elimination rules **selectors**. It is possible to give typed rules also for the modification of natural deduction of Section 3.4, the one with elimination rules of a general form, which leads to a generalized notion of application in lambda-calculus.

We can look at the proof terms of typed lambda-calculus as linear codifications of proof trees, conceived as algorithms that transform proofs of the open assumptions into a proof of the conclusion. Take as an example some neat natural derivation, such as the very first one in Chapter 3:

$$\dfrac{\dfrac{\dfrac{(A \supset B) \,\&\, (B \supset C)}{B \supset C}\,\&E_2 \qquad \dfrac{\dfrac{(A \supset B) \,\&\, (B \supset C)}{A \supset B}\,\&E_1 \qquad \overset{1}{A}}{B}\supset E}{C}\supset E}{A \supset C}\supset I,1$$

It is straightforward to add the proof terms to the natural derivation:

$$\dfrac{\dfrac{\dfrac{c : (A \supset B) \,\&\, (B \supset C)}{q(c) : B \supset C}\,\&E_2 \qquad \dfrac{\dfrac{c : (A \supset B) \,\&\, (B \supset C)}{p(c) : A \supset B}\,\&E_1 \qquad \overset{1}{w : A}}{app(p(c), w) : B}\supset E}{app(q(c), app(p(c), w)) : C}\supset E}{\lambda w \, app(q(c), app(p(c), w)) : A \supset C}\supset I,1$$

Whenever the schematic propositions A, B, and C have some concrete content, the proof transforms a categorical proof of $(A \supset B) \,\&\, (B \supset C)$ into a categorical proof of $A \supset C$. Here categorical means: with no open assumptions.

We can go one step further in the application of functional abstraction and close also the assumption $(A \supset B) \,\&\, (B \supset C)$, with the result:

$$\lambda c \, \lambda w \, app(q(c), app(p(c), w)) : (A \supset B) \,\&\, (B \supset C) \supset (A \supset C)$$

There are no open assumptions left, and the proof term is therefore closed, i.e., contains no free parameters. It does the job of the logical law of the transitivity of implication.

(b) Derivation and computation. In the natural representation of mathematical proofs, their characteristic form is that a claim B follows under some conditions A. This form is expressed concisely as the implication $A \supset B$ obtained by the rule of implication introduction. If at some stage the conditions A obtain, B follows by implication elimination. The Curry–Howard correspondence gives this latter step as a functional application: An argument $a : A$ is fed into the function $f : A \supset B$ and a value $app(f, a) : B$ obtained. Reasoning constructively, without the use of the classical law of excluded middle, the function $f : A \supset B$ is an algorithm, i.e., a computable function. Gentzen's idea of normalization has then the following specific meaning: Given the function $f : A \supset B$ and the argument $a : A$, normalization consists in the computation of the value of $app(f, a)$, into $f(a)$ obtained by substitution. The computation of the value is nothing but the conversion of the non-normal derivation into normal form, which makes clear the importance of **strong normalization** and **uniqueness.** The former concept requires that a normal form is reached independent of the order of conversions, the latter that the result moreover be unique. Let us make the above more precise:

Assume given $\lambda w\, b(w) : A \supset B$ and an actual proof of A, so $a : A$. The function $\lambda w\, b(w)$ is applied to a by substituting a for w in $b(w)$, with a proof of B as a value. A formal notation for the computation of the value is given by an equality in which the 'second-order' function app takes the function $\lambda w\, b(w)$ as a first argument and an argument of this latter function as a second argument, written altogether as $app(\lambda w\ b(w), a) = b[a/w]$. The value is thus obtained by the substitution $b[a/w]$. The step of natural deduction and its typed version are:

$$\dfrac{\dfrac{\begin{matrix}\overset{1}{A}\\ \vdots \\ B\end{matrix}}{A \supset B}\,{\scriptstyle \supset I,1} \qquad \begin{matrix}\vdots\\ A\end{matrix}}{B}\,{\scriptstyle \supset E} \qquad\qquad \dfrac{\dfrac{\begin{matrix}\overset{1}{w : A}\\ \vdots \\ b(w) : B\end{matrix}}{\lambda w\, b(w) : A \supset B}\,{\scriptstyle \supset I,1} \qquad \begin{matrix}\vdots\\ a : A\end{matrix}}{app(\lambda w\, b(w), a) : B}\,{\scriptstyle \supset E}$$

The natural derivation has a non-normality; the typed version has a selector term. The non-normality is eliminated by a detour conversion, the selector term by the computation given by the equation $app(\lambda w\ b(w), a) = b[a/w]$. It corresponds exactly to the detour conversion in which a proof of A is taken, continued by a proof of B from A:

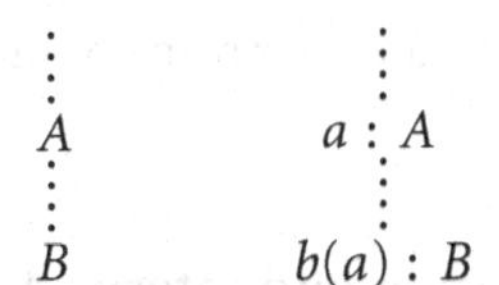

Conversion of proof terms works similarly in other cases of non-normality: Say, if we have proved $A \,\&\, B$ by $\&I$, then $pair(a, b) : A \,\&\, B$. An elimination rule with the conclusion A has the proof term $p(pair(a, b)) : A$. Normalization corresponds to the computation $p(pair(a, b)) = a$.

The rules for the quantifiers can be typed similarly to those for the connectives:

Table 13.5 Typed natural deduction rules for the quantifiers

$$\dfrac{a(x) : A(x)}{\lambda x\, a(x) : \forall x A(x)}\,{\scriptstyle \forall I} \qquad \dfrac{f : \forall x A(x)}{f(t) : A(t)}\,{\scriptstyle \forall E}$$

$$\dfrac{w : A(t)}{pair(t, w) : \exists x A(x)}\,{\scriptstyle \exists I} \qquad \dfrac{c : \exists x A(x) \quad \begin{matrix}\overset{1}{w : A(y)}\\ \vdots \\ a(w) : C\end{matrix}}{d(c, \lambda w\, a(w)) : C}\,{\scriptstyle \exists E,1}$$

In rule $\forall I$, the lambda-abstraction is taken over the individual eigenvariable x. The proof $a(x)$ is thus something that gives, for any individual term t, a proof of $a(t)$ through the substitution of t for x. Rule $\exists I$ has a term that gives an individual t and a proof w that t has the property $A(t)$. The typed quantifier rules lead to a remarkable unification in logic. When they are written as in Per Martin-Löf's **intuitionistic type theory** (1984), the following happens: In case A does not depend on x, i.e., is a constant proposition, the rules for $\forall$ are the same as those for $\supset$ and the rules for $\exists$ the same as those for &. In other words, implication turns out to be a special case of universal quantification and conjunction a special case of existential quantification.

Truth of a formula A is established by a proof $a : A$. Thus, *A is true* corresponds to A being, when considered as a set, *nonempty*. Typed lambda-calculus shows the rules of intuitionistic natural deduction to be correct (or, as one says in model theory, **sound**) under the computational semantics given by the Curry–Howard correspondence: If the premisses of a rule are assumed true, each of them has an element, and the rules show how to construct an element of the conclusion that thereby also must be true.

From the point of view of the Curry–Howard idea, formal proofs in intuitionistic natural deduction are computable functions. Constructivity, which used to be the philosophical principle behind intuitionistic logic and mathematics, now has the role of guaranteeing that computations do not go on indefinitely, but terminate after some bounded number of steps.

13.4 Cuts, their elimination and interpretation

Section 13.2 showed how two derivations in natural deduction can be combined into a single one:

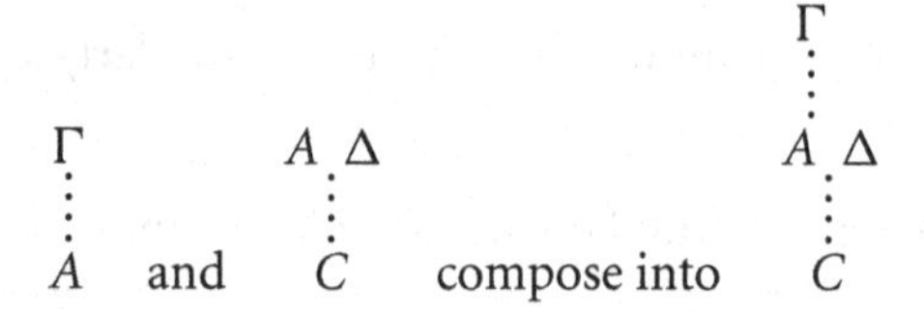

The condition on composition is that the discharge labels and eigenvariables in the two derivations to be composed are distinct. A proof of the admissibility of composition was likewise given in Section 13.2.

Composition corresponds to the **rule of cut** in sequent calculus. Gentzen's proof systems became first known mainly in the form of sequent calculi and the cut elimination theorem that we shall detail first. Then follows a precise determination of the correspondence of derivations in sequent calculus with cuts and derivations in natural deduction.

(a) Cut. Composition in natural deduction translates into what is called **cut** in sequent calculus:

$$\dfrac{\Gamma \vdash A \quad A, \Delta \vdash C}{\Gamma, \Delta \vdash C}\,\mathit{Cut}$$

Contrary to composition in natural deduction, the application of *Cut* in sequent calculus takes us outside the class of derivations by the logical rules for the connectives and quantifiers. It can be shown, instead, that if $\Gamma \vdash C$ is derivable by the logical rules and the rule of cut, it is derivable without the latter. Derivations can be so transformed that instances of *Cut* get eliminated one after the other. This result is often called the *Hauptsatz* (main theorem), the name Gentzen gave to it in 1933. It is analogous to normalization in a very precise sense: Any non-normal derivation can be translated into a sequent calculus derivation in which cuts are found in exactly those places in which there were non-normalities in the natural derivation. When these cuts are eliminated, the resulting cut-free derivation in sequent calculus can be translated back to a natural derivation that is always normal. For an example, consider the non-normality and its translation into sequent calculus:

$$\dfrac{\dfrac{\begin{matrix}\Gamma \\ \vdots \\ A\end{matrix} \quad \begin{matrix}\Delta \\ \vdots \\ B\end{matrix}}{A \,\&\, B}\,\&I \qquad \begin{matrix}\overset{1}{A}\ \overset{1}{B}\ \Theta \\ \vdots \\ C\end{matrix}}{C}\,\&E,1 \qquad\qquad \dfrac{\dfrac{\Gamma \vdash A \quad \Delta \vdash B}{\Gamma, \Delta \vdash A \,\&\, B}\,R\& \qquad \dfrac{A, B, \Theta \vdash C}{A \,\&\, B, \Theta \vdash C}\,L\&}{\Gamma, \Delta, \Theta \vdash C}\,\mathit{Cut}$$

The most central feature of cut elimination can now be illustrated: Instances of *Cut* are permuted up in a derivation, as in the transformation of the above sequent derivation into:

$$\dfrac{\Delta \vdash B \qquad \dfrac{\Gamma \vdash A \quad A, B, \Theta \vdash C}{B, \Gamma, \Theta \vdash C}\,\mathit{Cut}}{\Gamma, \Delta, \Theta \vdash C}\,\mathit{Cut}$$

A cut with the **cut formula** $A \,\&\, B$ is replaced by two cuts on the shorter cut formulas A and B. The rules of natural deduction lead to corresponding sequent calculus rules with independent contexts. After cut elimination, the translation back to natural deduction is done as in Table 4.4.

The details of cut elimination depend on the sequent calculus adopted. The following calculus has turned out to offer the best control over the structure of derivations: The basis is the intuitionistic sequent calculus with the logical rules of Table 4.1, and Table 9.4 for the quantifiers. Initial sequents must have the form $P, \Gamma \vdash P$ with P an atomic formula. (It is not difficult to prove that sequents of the form $A, \Gamma \vdash A$ with A an arbitrary formula, are then derivable.) Next to the logical rules, certain 'structural rules' are added, to be used in cut elimination:

Table 13.6 The rules of weakening and contraction

$$\frac{\Gamma \vdash C}{A, \Gamma \vdash C}\,Wk \qquad \frac{A, A, \Gamma \vdash C}{A, \Gamma \vdash C}\,Ctr$$

Cut is a rule with independent contexts, whereas the logical rules have shared contexts. In cut elimination, there is often a need to create shared contexts through weakening, or to remove duplications through contraction.

The proof of cut elimination is organized as follows:

1. It is shown that rule *Wk* is admissible, with the further property that the height of a derivation is preserved. In other words, if $\Gamma \vdash C$ is derivable with a height of derivation n, also $A, \Gamma \vdash C$ is derivable with a height of derivation n. The proof is easy: Just add the formula A to the antecedent of each sequent in the derivation. (If A contains an eigenvariable of the original derivation, change the latter.)

2. Rule *Ctr* is shown admissible, with the same property of preservation of derivation height as in rule *Wk*. The proof is not as straightforward as that for weakening (cf. *Structural Proof Theory*, p. 33 for details).

3. A derivation with cuts is considered, starting with a subderivation that ends with an uppermost instance of *Cut*. As long as the cut formula is not principal in the right premiss of *Cut*, it is permuted up. Let the part of derivation be:

$$\frac{\dfrac{\Gamma \vdash A \quad \Gamma \vdash B}{\Gamma \vdash A \& B}\,R\& \qquad \dfrac{A \& B, \Theta \vdash C}{A \& B, \Theta \vdash C \vee D}\,R\vee}{\Gamma, \Theta \vdash C \vee D}\,Cut$$

The part of derivation with cut permuted up is:

$$\frac{\dfrac{\dfrac{\Gamma \vdash A \quad \Gamma \vdash B}{\Gamma \vdash A \& B}\,R\& \qquad A \& B, \Theta \vdash C}{\Gamma, \Theta \vdash C}\,Cut}{\Gamma, \Theta \vdash C \vee D}\,R\vee$$

In the end, it can happen that $A \,\&\, B, \Theta \vdash C$ is derived by $L\&$ with $A \,\&\, B$ as the principal formula and cut is permuted as in the first example in which $A \,\&\, B$ was principal in both premisses of cut. Otherwise, the right premiss is an initial sequent and we have a step of the form:

$$\dfrac{\dfrac{\Gamma \vdash A \quad \Gamma \vdash B}{\Gamma \vdash A \,\&\, B}{\scriptstyle R\&} \qquad A \,\&\, B, P, \Theta \vdash P}{P, \Gamma, \Theta \vdash P}{\scriptstyle Cut}$$

The conclusion is an initial sequent and the part of derivation above it can be deleted.

If the cut formula is not principal in the left premiss of cut, the cut is permuted up similarly.

The inductive parameters in cut elimination are the number of cuts in a given derivation, the length of an uppermost cut formula to be eliminated, and finally the sum of the heights of derivation of the premisses of such an uppermost cut.

As said, the details of cut elimination depend on the kind of sequent calculus one uses. For example, Gentzen's original calculus had rules of weakening and contraction that were not eliminated. There were lots of cases in which, after cut was permuted up one step, weakenings and contractions had to be introduced to recreate the exact sequent that was concluded in the original cut.

Looking at the rules of sequent calculus in the direction of the conclusion, the structural ones included, we notice that cut is the only rule in which a formula disappears. Thus, moving in the opposite root-first direction, we could try to find derivations by instantiating cuts with ever new cut formulas, and proof search would not terminate. If contraction is an explicit part of a system of proof, proof search could similarly go on for ever, through the production of duplications by the rule of contraction. For this reason, the contraction-free sequent calculus is a clear improvement over that of Gentzen.

What has been said of the intuitionistic calculus holds also for the classical, symmetric sequent calculus. The rules of weakening and contraction apply on both sides of sequents:

Table 13.7 Weakening and contraction for symmetric sequents

$\dfrac{\Gamma \vdash \Delta}{A, \Gamma \vdash \Delta}{\scriptstyle LW}$	$\dfrac{\Gamma \vdash \Delta}{\Gamma \vdash \Delta, A}{\scriptstyle RW}$	$\dfrac{A, A, \Gamma \vdash \Delta}{A, \Gamma \vdash \Delta}{\scriptstyle LC}$	$\dfrac{\Gamma \vdash \Delta, A, A}{\Gamma \vdash \Delta, A}{\scriptstyle RC}$

With several formulas in the succedent part, there are permutations symmetric to those in which some other formula than the cut formula was principal in the antecedent of the right premiss.

(b) The interpretation of cuts in natural deduction. Right from the beginning of structural proof theory, it was clear that the introduction rules of natural deduction correspond to the right rules of sequent calculus and the elimination rules to the left ones. This much is made clear in Gentzen's dissertation that contains a translation from natural to sequent derivations. Oddly enough, cuts are inserted there each time an elimination rule is translated, even if the given natural derivation was normal.

In von Plato (2001), a translation was defined from normal derivations in natural deduction with general elimination rules such that no cuts were inserted. Moreover, the correspondence was isomorphic in the sense that it maintained the order of the logical rules. A non-normal instance of an E-rule became translated into a left rule followed immediately by a cut on the principal formula. Such cuts, in turn, could be interpreted as non-normalities, but the question remains how the rest of the cuts are to be interpreted in terms of natural deduction.

We begin by a translation from natural to sequent derivations, then determine what a reverse translation requires.

Step 1. Natural deduction in sequent calculus style. Natural derivations are written in sequent calculus style, as in the calculus **NLI** of Section 13.2.

Vacuous and **multiple** closing of assumptions can be treated in two ways, first, implicitly as in standard natural deduction, and exemplified by:

$$\dfrac{\begin{matrix}\vdots\\ A^n, \Gamma \vdash B\end{matrix}}{\Gamma \vdash A \supset B}\,R\supset$$

The case of $n = 0$ corresponds to weakening, $n > 1$ to contraction in sequent calculus. A proof of cut elimination for such a sequent calculus is found in *Structural Proof Theory*, chapter 5.

A second possibility is to use explicit rules of weakening and contraction, as in:

$$\dfrac{\begin{matrix}\Gamma\\ \vdots\\ B\end{matrix}}{A \supset B}\,\supset I \quad \rightsquigarrow \quad \dfrac{\dfrac{\begin{matrix}\vdots\\ \Gamma \vdash B\end{matrix}}{A, \Gamma \vdash B}\,Wk}{\Gamma \vdash A \supset B}\,R\supset \qquad \dfrac{\begin{matrix}\overset{1}{A}, \overset{1}{A}, \Gamma\\ \vdots\\ B\end{matrix}}{A \supset B}\,\supset I,1 \quad \rightsquigarrow \quad \dfrac{\dfrac{\begin{matrix}\vdots\\ A, A, \Gamma \vdash B\end{matrix}}{A, \Gamma \vdash B}\,Ctr}{\Gamma \vdash A \supset B}\,R\supset$$

We can now mostly leave weakening and contraction aside and assume that each closing of assumptions is simple (i.e., $n = 1$).

Elimination rules in sequent calculus style have the form:

$$\dfrac{\begin{matrix}\Gamma \\ \vdots \\ A \,\&\, B\end{matrix} \qquad \begin{matrix}\overset{1}{A}, \overset{1}{B}, \Delta \\ \vdots \\ C\end{matrix}}{C}\,{\scriptstyle \&E,1} \quad \rightsquigarrow \quad \dfrac{\begin{matrix}\vdots \\ \Gamma \vdash A \,\&\, B\end{matrix} \qquad \begin{matrix}\vdots \\ A, B, \Delta \vdash C\end{matrix}}{\Gamma, \Delta \vdash C}\,{\scriptstyle \&E}$$

Step 2. Translation to sequent calculus, normal E-rules. Step 1 gave for an E-rule, normal instance, the form:

$$\dfrac{A \,\&\, B \qquad \begin{matrix}\overset{1}{A}, \overset{1}{B}, \Gamma \\ \vdots \\ C\end{matrix}}{C}\,{\scriptstyle \&E,1} \quad \rightsquigarrow \quad \dfrac{A \,\&\, B \vdash A \,\&\, B \qquad \begin{matrix}\vdots \\ A, B, \Delta \vdash C\end{matrix}}{A \,\&\, B, \Delta \vdash C}\,{\scriptstyle \&E}$$

Next leave out the *MP* $A \,\&\, B \vdash A \,\&\, B$ and out comes a sequent rule:

$$\dfrac{\begin{matrix}\vdots \\ A, B, \Delta \vdash C\end{matrix}}{A \,\&\, B, \Delta \vdash C}\,{\scriptstyle L\&}$$

The rest of the E-rules are translated in exactly the same way, with the result:

Theorem 13.6. Isomorphic translation. *The translation of a normal derivation in natural deduction with general elimination rules gives a sequent derivation that is cut free and has the same order of logical rules.*

The translation goes both ways, so we get:

Corollary 13.7. Interpretation of cut-free derivations in natural deduction. *Sequent calculus is a method for constraining proof search to produce normal derivations in* **NLI**.

Step 3. Translation to sequent calculus, non-normal E-rules. Step 1 for a non-normal instance of an E-rule gave:

$$\dfrac{\begin{matrix}\Gamma \\ \vdots \\ A \,\&\, B\end{matrix} \qquad \begin{matrix}\overset{1}{A}, \overset{1}{B}, \Delta \\ \vdots \\ C\end{matrix}}{C}\,{\scriptstyle \&E,1} \quad \rightsquigarrow \quad \dfrac{\begin{matrix}\vdots \\ \Gamma \vdash A \,\&\, B\end{matrix} \qquad \begin{matrix}\vdots \\ A, B, \Delta \vdash C\end{matrix}}{\Gamma, \Delta \vdash C}\,{\scriptstyle \&E}$$

The third step of translation is:

$$\dfrac{\Gamma \vdash A \,\&\, B \qquad A, B, \Delta \vdash C}{\Gamma, \Delta \vdash C}\,{\scriptstyle \&E} \quad \rightsquigarrow \quad \dfrac{\Gamma \vdash A \,\&\, B \qquad \dfrac{A, B, \Delta \vdash C}{A \,\&\, B, \Delta \vdash C}\,{\scriptstyle L\&}}{\Gamma, \Delta \vdash C}\,{\scriptstyle Cut}$$

It is seen clearly that natural deduction, the inference at left, is closed with respect to non-normal rule instances, whereas sequent calculus, the inference at right, requires a cut for the corresponding inference.

Theorem 13.8. Isomorphic translation of non-normal derivations. *The translation of a non-normal derivation in natural deduction with general elimination rules gives a sequent derivation in which each derived MP of an E-rule is the principal formula of a left rule and a cut formula immediately after the left rule, and no other cut formulas appear.*

Again, the translation goes both ways, so that we can interpret certain cuts in terms of natural deduction, as non-normalities.

Translation of normal instances is in fact a special case of the above:

$$\frac{A\&B \vdash A\&B \quad A, B, \Delta \vdash C}{A\&B, \Delta \vdash C}{\scriptstyle \&E} \quad \rightsquigarrow \quad \frac{A\&B \vdash A\&B \quad \dfrac{A, B, \Delta \vdash C}{A\&B, \Delta \vdash C}{\scriptstyle L\&}}{A\&B, \Delta \vdash C}{\scriptstyle Cut}$$

Here we have the base case of the inductive procedure of cut elimination: The left premiss of *Cut* is an initial sequent so that the conclusion of *Cut* is identical to the right premiss and the rule instance can be deleted.

Let us summarize the results of the four translations:

1. *Normal to cut-free derivations*
2. *Cut-free to normal derivations*
3. *Non-normal derivations to derivations with* L*-rule*+Cut

The reverse of 3 gives a translation to natural deduction such that each *L*-rule+*Cut* turns into a non-normality, so we have:

4. *To non-normal derivations when all cut formulas in the right premiss of cut have been derived by a left rule*

(c) How to interpret arbitrary cuts? Something has to be added to natural deduction; the key is how two derivations are composed together. There is a gap in the standard notation for natural deduction, for it cannot be shown *when, if at all*, a composition was made. Here is a simple example:

$$\frac{A\&B \quad \overset{1}{B}}{B}{\scriptstyle \&E,1} \quad \text{and} \quad \dfrac{\dfrac{B}{B \vee C}{\scriptstyle \vee I}}{(B \vee C) \vee D}{\scriptstyle \vee I} \quad \text{compose into} \quad \dfrac{\dfrac{\dfrac{A\&B \quad \overset{1}{B}}{B}{\scriptstyle \&E,1}}{B \vee C}{\scriptstyle \vee I}}{(B \vee C) \vee D}{\scriptstyle \vee I}$$

No trace is left of the composition. In sequent notation, instead, we have:

$$\dfrac{\dfrac{A\,\&\,B \vdash A\,\&\,B \quad B \vdash B}{A\,\&\,B \vdash B}{\scriptstyle \&E} \qquad \dfrac{\dfrac{B \vdash B}{B \vdash B \lor C}{\scriptstyle \lor I}}{B \vdash (B \lor C) \lor D}{\scriptstyle \lor I}}{A\,\&\,B \vdash (B \lor C) \lor D}{\scriptstyle Comp}$$

When we translate a non-normal derivation into sequent calculus, the derived major premiss is at once cut from the derivation. In sequent calculus, instead, it is possible to cut it out at a later stage. Thus, what can be called **delayed composition** is possible in sequent calculus. This will be our interpretation in natural deduction of cuts that do not turn into non-normalities in translation. The rule of composition has to be made explicit to make such delays visible.

Main theorem 13.9. Interpretation of cuts in natural deduction. *Derivations in sequent calculus with cuts can be interpreted as derivations in natural deduction with explicit composition:*

1. *If the cut formula D is* **principal** *in rule $L\circ$ in the right premiss of cut, the cut is translated into a non-normal instance of rule $\circ E$ with derived major premiss D.*

2. *If the cut formula D is* **not principal**, *the cut is translated into an explicit composition.*

The first case is illustrated by the example:

$$\dfrac{\Gamma \vdash A\,\&\,B \quad \dfrac{A, B, \Delta \vdash C}{A\,\&\,B, \Delta \vdash C}{\scriptstyle L\&}}{\Gamma, \Delta \vdash C}{\scriptstyle Cut} \quad \rightsquigarrow \quad \dfrac{\Gamma \vdash A\,\&\,B \quad A, B, \Delta \vdash C}{\Gamma, \Delta \vdash C}{\scriptstyle \&E}$$

For the second case, it will be instructive to look at the elimination of *Comp* in detail. *Comp* will in general permute up just like that, as in:

$$\dfrac{\Gamma \vdash D \quad \dfrac{D, \Delta' \vdash C'}{D, \Delta \vdash C}{\scriptstyle I\text{-}Rule}}{\Gamma, \Delta \vdash C}{\scriptstyle Comp} \quad \rightsquigarrow \quad \dfrac{\dfrac{\Gamma \vdash D \quad D, \Delta' \vdash C'}{\Gamma, \Delta' \vdash C'}{\scriptstyle Comp}}{\Gamma, \Delta \vdash C}{\scriptstyle I\text{-}Rule}$$

Eventually either the base case of *Comp*-elimination is met and *Comp* disappears, else *Comp* hits an elimination rule.

Case of *Comp* with an E-rule:

$$\dfrac{\Gamma \vdash D \qquad \dfrac{(D), \Delta \vdash A\,\&\,B \quad (D), A, B, \Theta \vdash C}{D, \Delta, \Theta \vdash C}{\scriptstyle \&E}}{\Gamma, \Delta, \Theta \vdash C}{\scriptstyle Comp}$$

Comp is permuted up to whichever premiss contains D.

Crucial observation. *The permutations are trivial unless a normal instance of an E-rule is met.*

Permuting *Comp* up, it can happen that it meets a normal instance of an E-rule:

$$\cfrac{\Gamma \vdash A \& B \qquad \cfrac{A \& B \vdash A \& B \quad A, B, \Delta \vdash C}{A \& B, \Delta \vdash C}\,{\scriptstyle \&E}}{\Gamma, \Delta \vdash C}\,{\scriptstyle Comp}$$

Permuting *Comp* up at right one more step, we get:

$$\cfrac{\cfrac{\Gamma \vdash A \& B \quad A \& B \vdash A \& B}{\Gamma \vdash A \& B}\,{\scriptstyle Comp} \qquad A, B, \Delta \vdash C}{\Gamma, \Delta \vdash C}\,{\scriptstyle \&E}$$

This is the base case of elimination of *Comp*, so we get a non-normality with no *Comp* left:

$$\cfrac{\Gamma \vdash A \& B \quad A, B, \Delta \vdash C}{\Gamma, \Delta \vdash C}\,{\scriptstyle \&E}$$

Sequent calculus confounds cuts that correspond to non-normalities and cuts that correspond to compositions in natural deduction. An explicit rule of composition keeps track of when a composition was made, but *Comp* is not a rule of natural deduction in sequent calculus style. Closure with respect to *Comp* has to be shown for just one instance of *Comp*, and repeated instances of *Comp* are eliminated one at a time with an uppermost one first.

Two successive instances of *Comp* on C and D, respectively, can be permuted, as in:

$$\cfrac{\cfrac{\Gamma \vdash C \quad C, \Delta \vdash D}{\Gamma, \Delta \vdash D}\,{\scriptstyle CompC} \qquad D, \Theta \vdash E}{\Gamma, \Delta, \Theta \vdash E}\,{\scriptstyle CompD}$$

$$\rightsquigarrow \quad \cfrac{\Gamma \vdash C \qquad \cfrac{C, \Delta \vdash D \quad D, \Theta \vdash E}{C, \Delta, \Theta \vdash D}\,{\scriptstyle CompD}}{\Gamma, \Delta, \Theta \vdash E}\,{\scriptstyle CompC}$$

Just as with two cuts, the latter order of composition can be permuted back, with such back and forth repeated any number of times, so some say there is 'failure of strong cut elimination'.

Corollary 13.10 **The proper analogue of strong normalization in sequent calculus.** *An isomorphic image of normalization and strong normalization is projected on sequent derivations with cuts by the following procedure:*

1. *Translate a given derivation with arbitrary cuts into natural deduction in sequent calculus style* +Comp.

2. *Eliminate all instances of* Comp, *starting with uppermost ones.*

3. *Normalize as you please.*

4. *All major premisses of E-rules are of the form $D \vdash D$, and when they are left unwritten, the result is a cut-free derivation in sequent calculus. It is as unique as the result of normalization in 3.*

In 1, some cuts can vanish into non-normalities, and others turn to explicit instances of rule *Comp*. In 2, some non-normalities may be created in the elimination of *Comp*.

14 | Deductive machinery from Aristotle to Heyting

We shall review the development of logic, beginning with Aristotle's syllogistic logic. Next the tradition of algebraic logic is described, through the work of George Boole, Ernst Schröder, and Thoralf Skolem. There follows a section on axiomatic logic with two phases, the early work of Gottlob Frege, Giuseppe Peano, and Bertrand Russell, and a second phase with David Hilbert and Paul Bernays, and up to Heyting's intuitionistic logic in 1930. That is the point right before Gentzen's development of natural deduction and sequent calculus. The emphasis is on how the logical systems work, i.e., on their deductive machinery.

14.1 Aristotle's deductive logic

Aristotle's system of deductive logic, also known as the 'theory of syllogisms', has been interpreted in various ways in the long time since it was conceived. The situation is not different from the reading of other chapters of the formal sciences of antiquity, such as Euclid's geometry and works of Archimedes. When Frege invented predicate logic, he finished the presentation proudly with a reconstruction of the Aristotelian forms of propositions, such as *Every A is B* that is interpreted as $\forall x(A(x) \supset B(x))$, with a universal quantification over some domain and the predicates A and B. Frege reproduced similarly Aristotelian inferences, such as the conclusion *Every A is C* obtained from the premisses *Every A is B* and *Every B is C*, in the way shown in Section 9.1. Frege's interpretation has become the most common one, but we can also consider Aristotle's logic in itself, without such interpretations, and see that it works to perfection. Passages in Aristotle are referred to by the 'Bekker numbering' of his works.

(a) The forms of propositions. Aristotle's system of deductive logic is presented in his book *Prior Analytics*. It begins with four forms of propositions with a **subject** A and a **predicate** B:

Table 14.1 The Aristotelian forms of propositions

Universal affirmative:	*Every A is B*
Universal negative:	*No A is B*
Particular affirmative:	*Some A is B*
Particular negative:	*Some A is not-B*

There is also an **indefinite** form of proposition, *A is B*, not usually present in the rules of inference, even though it can so be (cf. *Prior Analytics*, 26a28).

Subjects and predicates together are **terms**. The indefinite form *A is B* has various other readings: *Subject A has the predicate B, predicate B belongs to subject A, B belongs to A*, etc. The last one is preferred by Aristotle, and he writes the other forms similarly:

B belongs to every A, B belongs to no A, B belongs to some A, B does not belong to some A.

Here the copula is written as one connected expression between the predicate and the subject, which underlines the formal character of the sentence construction.

A second reading is given to *B does not belong to some A*, namely *B does not belong to all A* (in 24a17). Another useful way of expressing the Aristotelian propositions is:

Every A is B, No A is B, Some A is B, Not every A is B.

Now the indefinite form *A is B* is a constant part of the propositions and the varying quantifier structure is singled out. It is seen clearly that the first and last are opposites, and that the second and third are likewise opposites.

The main principle in the formation of propositions is that subjects and predicates are treated symmetrically in the universal and particular propositions: Whenever *Every A is B* is a proposition, also *Every B is A* is one, and similarly for the universal negative and the particular forms. A formal structure is imposed that is not a natural feature of natural language, as in *Some man is wise*, the **converse** of which, *Some wise is a man*, would not be a natural expression, but would have to be paraphrased, as in *Some wise being is a man.*

We shall use the following notation for the propositions of Aristotle's logic:

Table 14.2 Notation for the Aristotelian propositions

$\Pi^+(A, B)$	for	*Every A is B*
$\Pi^-(A, B)$	for	*No A is B*
$\Sigma^+(A, B)$	for	*Some A is B*
$\Sigma^-(A, B)$	for	*Some A is not-B*

The **universal quantifier** of the Aristotelian form $\Pi^+(A, B)$ is explained as follows in the *Prior Analytics* (24b28):

A thing is said of all of another when there is nothing to be taken of which the other could not be said.

Aristotle is saying that universality means the lack of a counterexample, a common idea in logic ever since. The other forms of quantification are explained similarly and used in a justification of the Aristotelian rules. However, we do not need to go into the details, because it will turn out sufficient to treat the four Aristotelian forms as just atomic formulas with two terms but no further internal structure.

(b) Indirect proof. The two pairs $\Sigma^+(A, B)$, $\Pi^-(A, B)$ and $\Pi^+(A, B)$, $\Sigma^-(A, B)$ form between themselves **contradictory opposites**. Furthermore, because from $\Pi^-(A, B)$ the weaker $\Sigma^-(A, B)$ follows, also $\Pi^+(A, B)$ and $\Pi^-(A, B)$ together lead to a contradictory pair. We shall indicate the contradictory opposite of a formula P by the orthogonality symbol, $P^\perp$. (Note that $P^{\perp\perp}$ is identical to P.) In general, if an assumption P has led to contradictory consequences Q and $Q^\perp$, $P^\perp$ can be concluded and the assumption P **closed**. The rule of indirect proof thus takes on the following schematic form:

Table 14.3 The scheme of indirect proof

$$\dfrac{\begin{matrix} \overset{1}{P^m} & \overset{1}{P^n} \\ \vdots & \vdots \\ Q & Q^\perp \end{matrix}}{P^\perp}\ {\scriptstyle RAA,1}$$

This schematic proof figure is to be understood as follows: The assumption P may appear among those that were used in the derivations of Q and $Q^{\perp}$, respectively. Any numbers $m, n \geqslant 0$ of occurrences of P in the two **subderivations** can be closed at the inference. The closed ones are indicated by a suitable label, such as a number, so that each instance of rule *RAA* (for *reductio ad absurdum*) clearly shows which occurrences of P are closed at the inference. It is typical of Aristotle's proofs that an assumption closed in indirect proof occurs just once, i.e., either $m = 1, n = 0$ or $m = 0, n = 1$. We then have one of:

Table 14.4 Aristotelian special cases of indirect proof

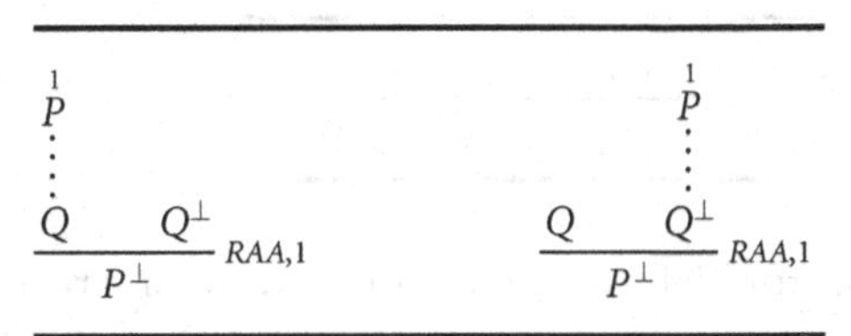

As mentioned, Aristotle's derivations have at most one instance of indirect proof, as a last rule. A rule of indirect proof in which the premisses of *RAA* are $\Pi^{+}(A, B)$ and its **contrary** $\Pi^{-}(A, B)$ can be derived from the second of the below conversion rules.

(c) The rules of conversion and syllogism. Aristotle's system of deductive logic begins properly with his **rules of conversion**:

Table 14.5 The rules of conversion

$$\frac{\Pi^{-}(A, B)}{\Pi^{-}(B, A)}\,\Pi^{-}C \qquad \frac{\Pi^{+}(A, B)}{\Sigma^{+}(B, A)}\,\Pi^{+}C \qquad \frac{\Sigma^{+}(A, B)}{\Sigma^{+}(B, A)}\,\Sigma^{+}C$$

The third rule of conversion is a **derivable rule**: Its conclusion is derivable from its premiss by the first conversion rule and the rule of indirect proof. Aristotle, in fact, notes the same (24a22).

Two more rules enter into Aristotle's deductive logic, the proper **syllogisms** as this word has been understood for a long time. Its meaning in Aristotle vacillates between a single syllogism and what today is called a deduction or derivation. The major part of the *Prior Analytics* deals with derivations that consist of a single syllogism, conversions, and a single step of indirect inference. The two syllogistic rules are (25b38–26a2):

Table 14.6 Aristotle's formulation of the syllogisms

When A of every B and B of every C, it is necessary that A is said of every C. For we have explained above what we mean by every.
Correspondingly also when A of no B, B instead of every C, then A will not belong to any C.

The added clause in the first syllogism hints at a justification of the rule in terms of the meaning given to universal quantification.

We write the above two rules as:

Table 14.7 The syllogistic rules

$$\frac{\Pi^{+}(A,B) \quad \Pi^{+}(B,C)}{\Pi^{+}(A,C)}\,\Pi^{+}S \qquad \frac{\Pi^{+}(A,B) \quad \Pi^{-}(B,C)}{\Pi^{-}(A,C)}\,\Pi^{-}S$$

The order of the premisses, from left to right, is the reverse of Aristotle's proof texts. At some stage, it became customary to read the propositions with the subject first, so, to have the middle term in the middle, the order of premisses was changed.

When one reads Aristotle's examples of syllogistic inference, the real deductive structure is somewhat hidden behind the convention of a linear sentence structure. Here is an example (from *Prior Analytics*, 27a10):

If M belongs to every N and to no X, then neither will N belong to any X. For if M belongs to no X, then neither does X belong to any M; but M belonged to every N; therefore, X will belong to no N (for the first figure has come about). And since the privative converts, neither will N belong to any X.

Let us number the sentences of this text in the succession in which they appear, rewritten so that each single purely syllogistic sentence is identified (i.e., with the connectives and rhetorical expressions eliminated):

1. *M belongs to every N.*
2. *M belongs to no X.*
3. *N belongs to no X.*
4. *M belongs to no X.*
5. *X belongs to no M.*
6. *M belongs to every N.*
7. *X belongs to no N.*
8. *N belongs to no X.*

The **assumptions** in the syllogistic proof are 1 and 2. Line 3 states the **conclusion** of the proof. Line 4 **repeats** assumption 2, and line 5 gives the result of application of a conversion rule to the premiss given by line 4. Line 6 repeats the assumption from line 1. Line 7 gives the conclusion of a syllogistic rule from the premisses given by lines 5 and 6. Line 8 gives the conclusion of a conversion rule applied to the premiss given by line 7. It is at the same time the sought-for conclusion expressed on line 3.

The formal nature of Aristotle's proof text is revealed by the repetition, twice, of assumptions or previous conclusions, on lines 4 and 6. These repetitions are made so that the application of a rule of inference in the proof text can follow a certain pattern: A one-premiss rule such as conversion is applied to a sentence in a way in which the conclusion follows immediately the sentence. A two-premiss rule such as a syllogism is applied to two premisses, given in succession in a predetermined order, so that the conclusion follows immediately the premisses.

(d) The deductive structure of syllogistic proofs. The linearity of Aristotle's proof texts hides a part of their true deductive structure. In the example, the two assumptions are **independent** of each other, neither is derivable from the other by the rules. Leaving out the tentative statement of the conclusion from line 3, the **deductive dependencies** on assumptions in Aristotle's proof are: Line 4 depends on assumption 2, line 5 likewise on 2 through 4, line 6 on 1, line 7 on 1 and 2, and line 8 on 1 and 2.

We shall translate Aristotle's linear derivation, with sentences numbered as in the example, into a tree form by the following two clauses, quite similar to those found in Chapter 3 when linear natural derivations were translated to tree form:

1. *Take the last sentence, draw a line above it, with the name of the rule that was used to conclude it next to the line. Write the sentences that correspond to the lines of the premisses of the last rule above the inference line.*
2. *With indirect proof, add a numerical label next to the rule.*
3. *Repeat the procedure until assumptions are arrived at. With indirect proof, add a numerical label on top of closed assumptions.*

Here is what we get when the translation algorithm is applied to the example text:

$$\dfrac{\dfrac{\dfrac{M\ \mathit{belongs\ to\ no}\ X}{X\ \mathit{belongs\ to\ no}\ M}\,{\scriptstyle Conv}\qquad M\ \mathit{belongs\ to\ every}\ N}{X\ \mathit{belongs\ to\ no}\ N}\,{\scriptstyle Syll}}{N\ \mathit{belongs\ to\ no}\ X}\,{\scriptstyle Conv}$$

The same derivation in our notation is:

$$\dfrac{\dfrac{\dfrac{\Pi^-(X, M)}{\Pi^-(M, X)}\,\Pi^-C \qquad \Pi^+(N, M)}{\Pi^-(N, X)}\,\Pi^-S}{\Pi^-(X, N)}\,\Pi^-C$$

As a next example, consider the following indirect syllogistic derivation (from 28b17):

If R belongs to every S but P does not belong to some, then it is necessary for P not to belong to some R. For if P belongs to every R and R to every S, then P will also belong to every S; But it did not belong.

Our notation and translation into tree form gives:

$$\dfrac{\dfrac{\overset{3}{\Pi^+(R, P)} \quad \Pi^+(S, R)}{\Pi^+(S, P)}\,\Pi^+S \qquad \Sigma^-(S, P)}{\Sigma^-(R, P)}\,RAA{,}3$$

To finish this brief tour of Aristotle's logic, we show that his practice of making at most one last step of indirect inference is justified:

Theorem 14.1. Normal form for derivations. *All derivations in Aristotle's deductive logic can be so transformed that the rule of indirect proof is applied at most once as a last rule.*

Proof. Consider an uppermost instance of *RAA* in a derivation. If it is followed by another instance of *RAA*, we have a part of derivation such as:

$$\dfrac{\dfrac{\begin{matrix}\overset{1}{P}\\ \vdots\\ Q\end{matrix} \quad \begin{matrix}\vdots\\ Q^{\perp}\end{matrix}}{P^{\perp}}\,RAA{,}1 \qquad \begin{matrix}\overset{2}{R}\\ \vdots\\ P\end{matrix}}{R^{\perp}}\,RAA{,}2$$

This derivation is transformed into:

$$\dfrac{\begin{matrix}\overset{1}{R}\\ \vdots\\ P\\ \vdots\\ Q\end{matrix} \quad \begin{matrix}\vdots\\ Q^{\perp}\end{matrix}}{R^{\perp}}\,RAA{,}1$$

Admissibility of composition guarantees that the derivation of P from R can be continued by the derivation of Q from P.

The above transformation is repeated until there is just one instance of *RAA*. If the conclusion $R^{\perp}$ is existential, it cannot be a premiss in any rule and the claim of the theorem follows. If the conclusion is universal, we have one of:

$$
\frac{\begin{array}{cc} \overset{1}{\Sigma^{-}(A,B)} & \\ \vdots & \vdots \\ Q & Q^{\perp} \end{array}}{\Pi^{+}(A,B)}\,{\scriptstyle RAA,1}
\qquad\qquad
\frac{\begin{array}{cc} \overset{1}{\Sigma^{+}(A,B)} & \\ \vdots & \vdots \\ Q & Q^{\perp} \end{array}}{\Pi^{-}(A,B)}\,{\scriptstyle RAA,1}
$$

There are by assumption no instances of *RAA* above the ones shown. Therefore the existential formulas that are closed cannot be premisses in any other rules than the instances of *RAA* shown. Then the derivations of the left premisses of *RAA* are degenerate, with $\Sigma^{-}(A, B) \equiv Q$ and $\Sigma^{+}(A, B) \equiv Q$, respectively. The derivations are therefore:

$$
\frac{\begin{array}{cc} & \vdots \\ \overset{1}{\Sigma^{-}(A,B)} & \Pi^{+}(A,B) \end{array}}{\Pi^{+}(A,B)}\,{\scriptstyle RAA,1}
\qquad\qquad
\frac{\begin{array}{cc} & \vdots \\ \overset{1}{\Sigma^{+}(A,B)} & \Pi^{-}(A,B) \end{array}}{\Pi^{-}(A,B)}\,{\scriptstyle RAA,1}
$$

A loop is produced in both, and therefore the instances of *RAA* can be removed. QED.

The formulas of Aristotle's deductive logic are atomic formulas in today's terminology, and his rules act only on such atomic formulas. The question of the derivability of an atomic formula from given atomic formulas used as assumptions is known as the **word problem**. The terminology stems from algebra where the word problem concerns the derivability of an equality from given equalities. The solution of this problem in Aristotle's deductive logic, i.e., the decidability of derivability by his rules, follows at once by the above result on normal form. It is sufficient to show that the terms in a derivation of P from the assumptions Γ can be restricted to those included in the assumptions and the conclusion. If the proof is direct, this is so because terms in any formula in the derivation can be traced up to assumptions. Otherwise, the last step is indirect, but the closed assumption is $P^{\perp}$ so that the terms in a derivation are again found in the assumptions or the conclusion P. With a bounded number of terms, there is a bounded number of distinct formulas. The number of possible consecutive steps of inference in a loop-free derivation, i.e., the height of a branch in a derivation tree, is bounded by the number of distinct formulas and we have:

Theorem 14.2. Word problem for Aristotle's deductive logic. *The derivability of a formula P from given formulas Γ used as assumptions is decidable.*

The *Prior Analytics* is rather brief on the general logical theory of syllogisms. Most of the treatise is devoted to modal syllogisms in which the basic form of expression is qualified, as in *A is possibly B* and *A is necessarily B*.

14.2 The algebraic tradition of logic

(a) Boole's logical algebra. Algebraic logic began in 1847 when George Boole presented his 'calculus of deductive reasoning' in a short book titled *The Mathematical Analysis of Logic*. His calculus reduced known ways of logical reasoning into the solution of algebraic equations. The known ways of logical reasoning were not just accounted for, but extended to full classical propositional logic.

Boole's starting point was Aristotle's theory of syllogistic inferences and its later development. Let us recall the propositions in Boole's notation:

Table 14.8 The four basic forms of syllogistic propositions

1. A: Each X is Y.
2. E: No X is Y.
3. I: Some X is Y.
4. O: Some X is not-Y.

Boole considered also more complicated forms, ones that appear in what are known as 'hypothetical syllogisms'.

Boole's logical calculus assumes as given a universe of objects, denoted 1. Classes of objects in the universe are denoted by $X, Y, Z, \dots$. Lower case letters $x, y, z, \dots$ are 'elective symbols' for these classes. The easiest way to understand these symbols is that x is a variable that takes the values 1 and 0 according to whether an object a belongs to the class X. The product xy is used for expressing that an object belongs to both X and Y, the sum $x + y$ for expressing that it belongs to at least one of X and Y, and the difference $1 - x$ for expressing that the object belongs to not-X, i.e., is a 'not-X' type of object.

The reading that is closest to Boole takes $X, Y, Z, \dots$ to be subsets of the universe of objects, with not-X given by the complement of a set, product by the intersection, and sum by the union of two sets. We can equivalently use monadic predicate logic, by which the predicates $X, Y, Z, \dots$ are applied to objects $a, b, c, \dots$ with elementary propositions such as $X(a)$ as results. Anachronistically speaking, Boole invented the valuation semantics of classical monadic predicate logic, therefore also of classical propositional logic, and this even before these two systems of logic had a well-defined syntax. The small variables represent valuations over the propositions: We have a valuation function v with $v(X(a)) = x$ and with $x = 1$ if $X(a)$ is true, $x = 0$ if $X(a)$ is false.

Boole writes down in an intuitive way properties of valuations, such as $xy = yx$ or $x(y + z) = xy + xz$. The syllogistic forms of Table 14.8 are represented in terms of equations as:

Table 14.9 Propositions represented as algebraic equations

1. A:	With X a subclass of Y, we have $xy = x$, so $x(1 - y) = 0$.
2. E:	If no X is Y, then $xy = 0$.
3. I:	With V the class of those X that are also Y, $v = xy$ gives I.
4. O:	Similarly to 3, we have $v = x(1 - y)$.

We are now ready to **reason by calculation**. Consider the first syllogistic inference in Aristotle. The premisses and their algebraic representations are:

Each X is Y, $\quad x(1 - y) = 0$,

Each Y is Z, $\quad y(1 - z) = 0$.

We have thus $x - xy = 0$ from the first premiss. Multiply $1 - z$ by the left side, to get $x(1 - z) - xy(1 - z) = 0$. Since by the second premiss $y(1 - z) = 0$, also $xy(1 - z) = 0$, so that $x(1 - z) = 0$. This just states that every X is Z, or the conclusion of the syllogism.

Boole moves to hypothetical propositions such as: If A is B, then C is D. It has the form: If X is true, then Y is true. The four possibilities, X true and Y true, X true and Y false, X false and Y true, X false and Y false, are represented by xy, $x(1 - y)$, $(1 - x)y$, and $(1 - x)(1 - y)$, respectively. It is now clear that any correct inference of classical propositional logic is validated by Boole's algebraic semantics, of which the 'truth tables' popularized by Wittgenstein are a notational variant. Thus, Boole has no difficulty to continue the list of 'principal forms of hypothetical Syllogism which logicians have recognized', as in his final example:

Table 14.10 Boole's final example of a hypothetical syllogism

If X is true, then either Y is true, or Z is true. But Y is not true. Therefore, if X is true, Z is true.

The disjunction in the first premiss is exclusive and can be given as the term $y + z - yz$. Then the first premiss is $x(1 - y - z + yz) = 0$. The second premiss is $y = 0$, and the conclusion $x(1 - z) = 0$ follows at once by easy calculation.

Boole did not put down any definitive list of algebraic laws that would define what we today call a Boolean algebra. The usual way to introduce such algebras is to start from classical propositional logic, then to collect all formulas equivalent to a given formula A into an equivalence class denoted $[A]$. Product, sum, and complement relative to the universe 1 in these classes correspond to the logical operations through the formation of the equivalence classes $[A\&B]$, $[A \vee B]$, and $[\neg A]$.

Boole reduced Aristotelian syllogistic reasoning to calculation, which was a wonderful achievement. Encouraged by the success, he wrote a book with the bold title *The Laws of Thought* (1854). His logic was not able to treat relations, but just one-place predicates, however. Today we know that there is no algebra of logic for full predicate logic, in which logical reasoning could be reduced to algebraic computation in the way Boole did for monadic predicate logic.

(b) Schröder's algebraic logic. The next important person in the development of the algebraic approach to logic and logic in general was Ernst Schröder. His work is found in the three-volume *Vorlesungen über die Algebra der Logik*, published between 1890 and 1905. He goes beyond Boole in that there is as a basic structure a partial order relation over objects, called 'groups', or 'domains' (*Gebiet*), with areas of the blackboard delimited by circles and ovals as a paradigmatic example. The order relation is used to represent logical consequence where Boole reasoned in terms of equalities. There are operations such as product and sum and relative complementation, with an obvious interpretation on the blackboard. Schröder's 'Gruppenkalkül' amounts to the study of logic in terms of lattice theory, though the latter terminology and its German equivalent 'Verbandstheorie' are of later origin. The partial order relation $a \leqslant b$ in a lattice has various readings, one of which is set inclusion, another logical consequence. Schröder's own symbol for the order is produced by superposing something like a subset symbol and an equality, the latter indicating that the 'subsumption' need not be strict, as in $\mathrel{\overline{\in}}$. The axioms are (Schröder (1890), pp. 168 and 170):

I. $a \mathrel{\overline{\in}} a$.
II. When $a \mathrel{\overline{\in}} b$ and $b \mathrel{\overline{\in}} c$ then also $a \mathrel{\overline{\in}} c$.

Algebraic laws determine unique lattice operations $a \cdot b$, also written ab, and $a + b$ (product and sum) that correspond to conjunction and disjunction in the logical reading. The principles that govern these operations are written as the 'definitions' (p. 197):

$(3_\times)'$ If $c \mathrel{\overline{\in}} a$, $c \mathrel{\overline{\in}} b$, then $c \mathrel{\overline{\in}} ab$.
$(3_\times)''$ If $c \mathrel{\overline{\in}} ab$, then $c \mathrel{\overline{\in}} a$ and $c \mathrel{\overline{\in}} b$.
$(3_+)'$ If $a \mathrel{\overline{\in}} c$, $b \mathrel{\overline{\in}} c$, then $a + b \mathrel{\overline{\in}} c$.
$(3_+)''$ If $a + b \mathrel{\overline{\in}} c$, then $a \mathrel{\overline{\in}} c$ and $b \mathrel{\overline{\in}} c$.

There are in addition two special domains 0 and 1 with $0 \mathrel{\overline{\in}} c$ and $c \mathrel{\overline{\in}} 1$. By setting c equal to ab in $(3_\times)''$, one obtains the standard lattice axioms $ab \mathrel{\overline{\in}} a$ and $ab \mathrel{\overline{\in}} b$. Axiom $(3_+)''$ gives similarly the axioms $a \mathrel{\overline{\in}} a + b$ and $b \mathrel{\overline{\in}} a + b$.

One famous problem in Schröder's logic concerns the **law of distributivity**. It is expressed in Schröder's language ((1890), p. 282) as the equation $a(b + c) = ab + ac$ with equality defined as partial order in both directions, and in a dual formulation as $(a + b)(a + c) = a + bc$. Is there a derivation of distributivity in Schröder's *Gruppenkalkül*? The founder of pragmatism, Charles Peirce, was one of the algebraic logicians and he believed himself to have proved the law. However, Schröder found a counterexample, with explicit reference to analogous counterexamples in Euclidean geometry: It consists of three circular areas a, b, c that intersect in a canonical way. With sum as union and product as intersection of the areas, it is readily seen that the dual formulation of distributivity fails in this case (p. 286). The subsumption relation goes only in one way, not two, as would be required by the definition of equality. Meanwhile also Peirce had come to recognize that his purported proof was fallacious (p. 290).

An abstract formulation of Schröder's counterexample to distributivity is given by a lattice that consists of just five distinct elements a, b, c, d, e with the following orderings:

$$d \leqslant b,\ d \leqslant a,\ d \leqslant c,\ b \leqslant e,\ a \leqslant e,\ c \leqslant e$$

A figure will be useful for computing the terms in the distributive inequality:

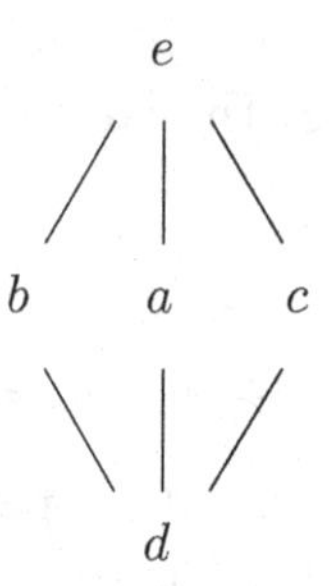

From the figure can be read off the equations $a(b+c) = ae = a$ as well as $(ab) + (ac) = dd = d$. For distributivity, we should have $a \leqslant d$, but $d \leqslant a$ was assumed. By the condition that all the elements be distinct, distributivity fails. This is the standard counterexample to distributivity in today's lattice theory.

The calculus of groups leads to a calculus of classes through the addition of an operation of negation, denoted $a_{|}$ and interpreted as a complement for areas. Schröder uses the algebraic calculus in the same way as Boole. The letters indicate properties and the task is to show what can be inferred from given letters used as assumptions. Here is one example (p. 530):

Let it be stipulated that every b that is not d is either both a and c or neither a nor c. Further, no c and no d can be a and b simultaneously. To prove that no a is b.

The assumptions are expressed as:

$$bd_{|} \mathrel{\underline{\in}} ac + a_{|}c_{|}\quad c \mathrel{\underline{\in}} (ab)_{|} \quad \text{and} \quad d \mathrel{\underline{\in}} (ab)_{|}$$

One of Schröder's basic observations is that the subsumption $a \mathrel{\underline{\in}} b$ is equivalent to the equalities $a = ub$ and $b = a+v$, for any u and v (p. 398). In particular, $a \mathrel{\underline{\in}} b$ whenever $ab_{|} = 0$. The three subsumptions that are assumed lead by this method into what is called a 'combined equation':

$$(ac_{|} + a_{|}c)bd_{|} + abc + abd = 0$$

Since $uu_{|} = 0$ and $u + u_{|} = 1$ for any u, the terms c and d can be eliminated from this equation, with the result $ab = 0$ as required. By the above, it follows that $a \mathrel{\underline{\in}} b_{|}$, i.e., that no a is b.

The example shows how logical consequence relations, as expressed by the subsumption relation, are turned into equalities on which algebraic manipulations in the style of Boole can be performed, to obtain a result that can be finally read again in terms of consequence.

(c) Skolem's combinatorics of deduction. Skolem's famous paper of 1920 contains the crowning achievement of algebraic logic. The paper, though, is not known for this, but for its first section that contains the Löwenheim–Skolem theorem, discussed in Section 10.2. The other parts were completely forgotten, together with the algebraic logic of Schröder that got transformed into lattice theory. Part of the reason was the notation: Skolem wrote (xy) for Schröder's $x ⋹ y$, the partial order relation $x \leqslant y$, and $\overset{\frown}{xyz}$ for the relation Schröder wrote as $xy = z$ and that we wrote as $M(x, y, z)$ ('the meet of x and y is z') in Section 11.3, and similarly $\underset{\smile}{xyz}$ for Schröder's $x+y = z$, the join relation $J(x, y, z)$ of Section 11.3.

Skolem wrote in 1913 a Master's thesis in Norwegian, titled *Undersökelser innenfor logikkens algebra* (Investigations in the algebra of logic). A part of the results was published in 1913 in the article *Om konstitutionen av den identiske kalkuls grupper* (On the structure of groups in the identity calculus). The paper begins with an admirably concise account of the algebraic approach to logic. The lattice ordering $a \leqslant b$ has various readings, one of which is set inclusion, another logical consequence. Algebraic laws determine unique lattice operations ab and $a + b$ (product and sum) that correspond to conjunction and disjunction in the logical reading. Negation $\overline{a}$ is defined by introducing a null class 0 and a universal class 1 satisfying an algebraic law that Skolem writes as:

$$(a\overline{a} \leqslant 0)(1 \leqslant a + \overline{a})$$

Now the lattice structure becomes Schröder's 'identity calculus', or a Boolean algebra in modern terms. Implication between a and b is later defined as the supremum of x such that $ax \leqslant b$.

In a later paper of 1919, Skolem studied further the notion of implication, or the case of a lattice in which the inequality $ax \leqslant b$ has a maximal solution, denoted by $\frac{b}{a}$ in Skolem. The resulting structure is in fact what is today called a **Heyting algebra**. The modern notation introduces an arrow operation $a \rightarrow b$, and Skolem's inequality written with the lattice meet and arrow operations reads as $a \wedge (a \rightarrow b) \leqslant b$. Many basic properties of Heyting algebras are proved, for which see von Plato (2007).

It is quite astonishing that Skolem had found an algebraic axiomatization of intuitionistic propositional logic well before its basic principles were definitively clarified by Heyting's axiomatization (1930). Of his motivations for introducing the algebraic axiomatization, what he called 'class rings' (*Klassenringe*), Skolem writes that they are 'a natural continuation and generalization of the groups of the identity calculus'.

Let us now turn to Skolem's 1920 paper. It gave the axioms of Schröder's lattice theory as **production rules**:

I. For each x, (xx).
II. From (xy) in combination with (yz) follows (xz).
$\text{III}_{\times}$. From $\overset{\frown}{xyz}$ follow (zx) and (zy).
III_{+}. From $\underset{\smile}{xyz}$ follow (xz) and (yz).

$\text{IV}_{\times}$. From $\overset{\frown}{xyz}$ in combination with (ux) and (uy) follows (uz).
IV_{+}. From $\underset{\smile}{xyz}$ in combination with (xu) and (yu) follows (zu).

$\text{V}_{\times}$. From $\overset{\frown}{xyz}$ in combination with (xx'), $(x'x)$, (yy'), $(y'y)$, (zz'), and $(z'z)$ follows $\overset{\frown}{x'y'z'}$.
V_{+}. From $\underset{\smile}{xyz}$ in combination with (xx'), $(x'x)$, (yy'), $(y'y)$, (zz'), and $(z'z)$ follows $\underset{\smile}{x'y'z'}$.

$\text{VI}_{\times}$. There is for arbitrary x and y a z such that $\overset{\frown}{xyz}$.
VI_{+}. There is for arbitrary x and y a z such that $\underset{\smile}{xyz}$.

Principles $V_{\times}$ and V_{+} contain assumptions such as (xx'), $(x'x)$. It would have helped the reader if Skolem had written out these as $x = x'$, etc., so that one could see that these two principles just state that equals can be substituted in the meet and join expressions. No such concessions are made to the reader. Had Skolem been used to the notation of *Principia Mathematica* by the time of the article, he might have written the production rules as logical axioms:

I. (xx).
II. $(xy) \,\&\, (yz) \supset (xz)$.
$\text{III}_{\times}$. $\overset{\frown}{xyz} \supset (zx)$, $\quad \overset{\frown}{xyz} \supset (zy)$.
$\text{IV}_{\times}$. $\overset{\frown}{xyz} \,\&\, (ux) \,\&\, (uy) \supset (uz)$.
$\vdots$
VI_{+}. $(x)(y)(\exists z)\, \underset{\smile}{xyz}$.

Even if there is no such symbolic language for the production rules, Skolem takes a purely formal and combinatorial view of them, one usually associated with Hilbert rather than the algebraic logicians:

> The validity of a sentence in algebra, based on this axiomatic foundation, consists simply in the possibility of proving the following: Given these and these pairs and triples (xy), $\overset{\frown}{xyz}$, etc., those and those pairs and triples can be derived by possibly repeated and combined applications of the axioms...

> In fact, *the axioms presented are production principles* by which new pairs and triples are derived from certain initial symbols . . .
>
> *Here we have a purely combinatorial conception of deduction on which I would like to put emphasis, because it proves to be especially useful in logical investigations.*

Skolem's paper has no formal notation for derivations, either. In proving his results, he writes things like 'consider one application of the principle of substitution of equals as a last step'. No trace is left of how he did his proofs that involve transformations of the order of application of the rules; in his head, on discarded paper? In my paper *In the shadows of the Löwenheim–Skolem theorem: Early combinatorial analyses of mathematical proofs* (2007) I show in great detail that Skolem's 1920 solution of the word problem for lattices is based on the permutation of the order of application of the production rules.

Let's look here at Skolem's example of a formal proof. It is the converse to the distributive law, expressed in the language of lattice theory with operations as $(a \wedge b) \vee (a \wedge c) \leqslant a \wedge (b \vee c)$. Skolem's proof text in the language and notation of his relational lattice theory is:

> From $\overset{\frown}{abd}$ in combination with $\overset{\frown}{ace}$, $\underset{\smile}{bcf}$, $\underset{\smile}{deg}$, and $\overset{\frown}{afh}$ follows the pair (gh). Proof: From $\underset{\smile}{bcf}$ follows by III_{+} the pair (bf). From $\overset{\frown}{abd}$ follow (da) as well as (db) by $\mathrm{III}_{\times}$. From (db) and (bf) follows by II (df). From (da) and (df) in combination with $\overset{\frown}{afh}$ follows by $\mathrm{IV}_{\times}$ [original has $\mathrm{VI}_{\times}$] (dh). From $\underset{\smile}{bcf}$ follows on the force of III_{+} the pair (cf). From $\overset{\frown}{ace}$ follow by $\mathrm{III}_{\times}$ (ea) and (ec). From (ec) and (cf) [original has (ef)] follows by II (ef). From (ea) in combination with (ef) and $\overset{\frown}{afh}$ follows by $\mathrm{IV}_{\times}$ [original has $\mathrm{VI}_{\times}$] (eh). From (dh) in combination with (eh) and $\underset{\smile}{deg}$ follows finally by IV_{+} (gh).

The proof is easier to read if we eliminate **linguistic variation**, often used to break the monotonicity of sentence construction in a proof text. The sentences in Skolem's proof are then, numbered and with the theorem to be proved added as line 0:

0. From $\overset{\frown}{abd}$, $\overset{\frown}{ace}$, $\underset{\smile}{bcf}$, $\underset{\smile}{deg}$, and $\overset{\frown}{afh}$ follows (gh).
1. From $\underset{\smile}{bcf}$ follows by III_{+} (bf).
2. From $\overset{\frown}{abd}$ follows by $\mathrm{III}_{\times}$ (da) and (db).
3. From (db) and (bf) follows by II (df).
4. From (da) and (df) and $\overset{\frown}{afh}$ follows by $\mathrm{IV}_{\times}$ (dh).
5. From $\underset{\smile}{bcf}$ follows by III_{+} (cf).
6. From $\overset{\frown}{ace}$ follows by $\mathrm{III}_{\times}$ (ea) and (ec).

7. From (ec) and (cf) follows by II (ef).
8. From (ea) and (ef) and $\overset{\frown}{afh}$ follows by $\text{IV}_\times$ (eh).
9. From (dh) and (eh) and $\underset{\smile}{deg}$ follows by IV_+ (gh).

Now eliminate the conjunctive conclusion from line 2:

2. From $\overset{\frown}{abd}$ follows by $\text{III}_\times$ (da) and (db)

becomes:

2. From $\overset{\frown}{abd}$ follows by $\text{III}_\times$ (da)

2′. From $\overset{\frown}{abd}$ follows by $\text{III}_\times$ (db)

Do similarly for line 6, and next write out the 'given pairs and triples' on lines that begin a proof, as in:

1. Given $\overset{\frown}{abd}$
2. Given $\overset{\frown}{ace}$
3. Given $\underset{\smile}{bcf}$
4. Given $\underset{\smile}{deg}$
5. Given $\overset{\frown}{afh}$

The result is, with some renumbering of lines and aligning:

1. Given	$\overset{\frown}{abd}$
2. Given	$\overset{\frown}{ace}$
3. Given	$\underset{\smile}{bcf}$
4. Given	$\underset{\smile}{deg}$
5. Given	$\overset{\frown}{afh}$
6. From $\underset{\smile}{bcf}$ follows by III_+	(bf)
7. From $\overset{\frown}{abd}$ follows by $\text{III}_\times$	(da)
8. From $\overset{\frown}{abd}$ follows by $\text{III}_\times$	(db)
9. From (db) and (bf) follows by II	(df)
10. From (da) and (df) and $\overset{\frown}{afh}$ follows by $\text{IV}_\times$	(dh)
11. From $\underset{\smile}{bcf}$ follows by III_+	(cf)
12. From $\overset{\frown}{ace}$ follows by $\text{III}_\times$	(ea)
13. From $\overset{\frown}{ace}$ follows by $\text{III}_\times$	(ec)
14. From (ec) and (cf) follows by II	(ef)
15. From (ea) and (ef) and $\overset{\frown}{afh}$ follows by $\text{IV}_\times$	(eh)
16. From (dh) and (eh) and $\underset{\smile}{deg}$ follows by IV_+	(gh)

The final step is to replace the premiss formulas by the numbers of their lines, to see better what depends on what. The result is:

1.	Given	$\overset{\frown}{abd}$
2.	Given	$\overset{\frown}{ace}$
3.	Given	$\underset{\smile}{bcf}$
4.	Given	$\underset{\smile}{deg}$
5.	Given	$\overset{\frown}{afh}$
6.	From 3 follows by III_+	(bf)
7.	From 1 follows by $\mathrm{III}_\times$	(da)
8.	From 1 follows by $\mathrm{III}_\times$	(db)
9.	From 8 and 6 follows by II	(df)
10.	From 7 and 9 and 5 follows by $\mathrm{IV}_\times$	(dh)
11.	From 3 follows by III_+	(cf)
12.	From 2 follows by $\mathrm{III}_\times$	(ea)
13.	From 2 follows by $\mathrm{III}_\times$	(ec)
14.	From 13 and 11 follows by II	(ef)
15.	From 12 and 14 and 5 follows by $\mathrm{IV}_\times$	(eh)
16.	From 10 and 15 and 4 follows by IV_+	(gh)

Next apply the following prescription:

1. *Write the last formula.*
2. *Draw a line above it.*
3. *Write next to the line the rule that was used.*
4. *Write above the line in order the premisses of the rule.*
5. *Repeat until you arrive at the given formulas.*

Translation of the production rules into tree form gives:

$$\frac{}{(xx)}\,I \qquad \frac{(xy)\quad(yz)}{(xz)}\,II \qquad \frac{\overset{\frown}{xyz}}{(zx)}\,III_\times \qquad \frac{\overset{\frown}{xyz}}{(zy)}\,III_\times \qquad \frac{(ux)\quad(uy)\quad\overset{\frown}{xyz}}{(uz)}\,IV_\times$$

With $x = u$ standing for (xu) and (ux), the rules of substitution can be written as:

$$\frac{x=u\quad y=v\quad z=w\quad \overset{\frown}{xyz}}{\overset{\frown}{uvw}}\,V_\times \qquad \frac{x=u\quad y=v\quad z=w\quad \underset{\smile}{xyz}}{\underset{\smile}{uvw}}\,V_+$$

Skolem's example derivation becomes translated into the tree:

$$\dfrac{\dfrac{\dfrac{\overset{\frown}{abd}}{(da)}{\scriptstyle III_\times} \qquad \dfrac{\dfrac{\overset{\frown}{abd}}{(db)}{\scriptstyle III_\times} \quad \dfrac{\underset{\smile}{bcf}}{(bf)}{\scriptstyle III_+}}{(df)}{\scriptstyle II} \qquad \overset{\frown}{afh}}{(dh)}{\scriptstyle IV_\times} \qquad \dfrac{\dfrac{\overset{\frown}{ace}}{(ea)}{\scriptstyle III_\times} \qquad \dfrac{\dfrac{\overset{\frown}{ace}}{(ec)}{\scriptstyle III_\times} \quad \dfrac{\underset{\smile}{bcf}}{(cf)}{\scriptstyle III_+}}{(ef)}{\scriptstyle II} \qquad \overset{\frown}{afh}}{(eh)}{\scriptstyle IV_\times} \qquad \underset{\smile}{deg}}{(gh)}{\scriptstyle IV_+}$$

Now we see the **deductive dependencies** in Skolem's proof:

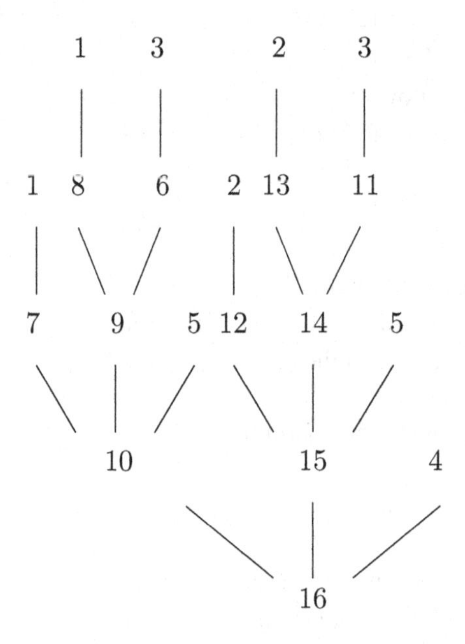

The traditional linear form of a written proof text hides these.

The following can be proved:

The rules of substitution permute down relative to the other rules and two consecutive instances of a rule of substitution contract into one.

Skolem may have thought such results too obvious to mention. The details can be found in *Proof Analysis*, section 5.3(b).

Skolem's treatment of lattice theory as a system of production rules ends with some examples of purely syntactic proofs of independence through failed proof search, including the distributive law, or what is written as $a \wedge (b \vee c) \leqslant (a \wedge b) \vee (a \wedge c)$ today:

Example 2. The task is to prove that the subsumption $a(b + c)$ € $ab + ac$ (in Schröder's notation), generally valid in the calculus of classes, is not generally valid in the calculus of groups. Translated into our language, this task means the following: Let the triples $\overset{\frown}{abd}$, $\overset{\frown}{ace}$, $\underset{\smile}{bcf}$, $\underset{\smile}{deg}$, and $\overset{\frown}{afh}$ be given. Does the pair (hg) follow from these by axioms I–VI? To investigate this, it suffices for us to apply axioms I–V as long as there appears a set S closed with respect to these axioms. Axiom I gives us the pairs (aa) (bb) (cc) (dd) (ee) (ff) (gg) (hh). Axiom III gives us (da) (db) (ea) (ec) (bf) (cf) (dg) (eg) (ha) (hf), and we get further by II (df) and (ef) and by IV_+ also (gf). Now, however, no more pairs or triples can be formed with the help of I–V from the 5 given triples. The pair (hg) does not appear among the pairs obtained. By this, the underivability of (hg) from the 5 given triples by axioms I–VI is proved or, in other words, the unprovability in the calculus of groups of what is known as the distributive law $a(b + c)$ € $ab + ac$.

The lattice-theoretic part is followed by another one on plane projective geometry, and a similar syntactic proof of the independence of what is known as Desargues' conjecture. Skolem's anticipation of proof search methods in algebra and geometry remained unnoticed for some eighty years – a lost opportunity in foundational research.

14.3 The logic of Frege, Peano, and Russell

(a) Frege's axiomatic predicate logic. Frege is the founding father of contemporary logic, through his little book *Begriffsschrift* that came out in 1879. The name stands for something like 'conceptual notation', and there is a long subtitle that specifies the notation as 'a formula language for pure thought, modeled upon that of arithmetic'. The actual notation in Frege's book is rather bizarre, and no one else has ever used it. Luckily he had Bertrand Russell among his few readers; Russell rewrote Frege's formula language in a style, adopted from Giuseppe Peano, that later evolved into the standard logical notation we have today.

The unique feature in Frege is that he wrote his formulas in two dimensions. The notation for an implication $A \supset B$ is:

This looks just like a vertical notation, but iterated implications show how it really is. Let us take as an example the formula that is written in standard notation as:

$$(C \supset (B \supset A)) \supset ((C \supset B) \supset (C \supset A))$$

Frege's writing is:

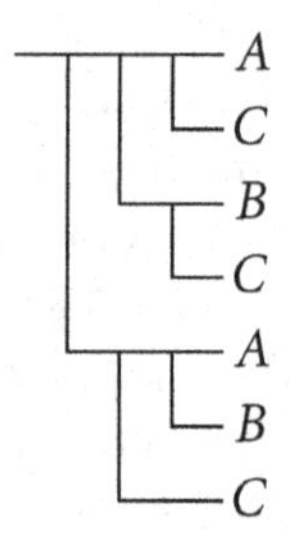

No parentheses are needed.

Next to implication, there is a negation sign that is just a little stroke, as in:

$$-\!\!\top\!\!- A$$

The universal quantifier is written so that the variable that is quantified is written in a little notch in the horizontal line at the head of a formula, as in:

$$-\!\!\overset{x}{\smile}\!\!- A(x)$$

Frege's logic is classical, and therefore implication and negation and the universal quantifier suffice for the definition of other connectives, such as conjunction and disjunction, and for the definition of existence.

Frege makes a careful distinction between a **proposition** and an **assertion**: From the former, the latter is obtained by the addition of the assertion symbol, a vertical line as in:

$$\vdash A$$

This notation has led to the turnstile that is used for the derivability relation.

The two-dimensional nature of Frege's formulas is best seen if we manipulate them a bit. Turn first the above formula 90 degrees:

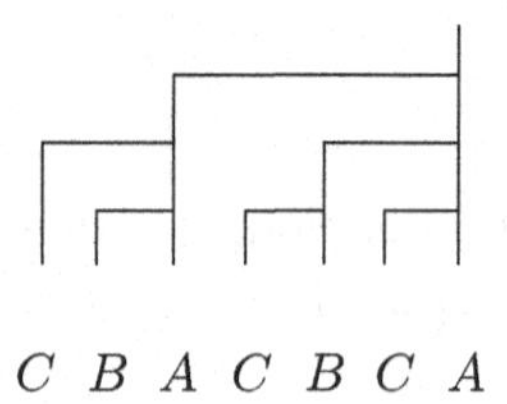

Straightening the lines, the result is:

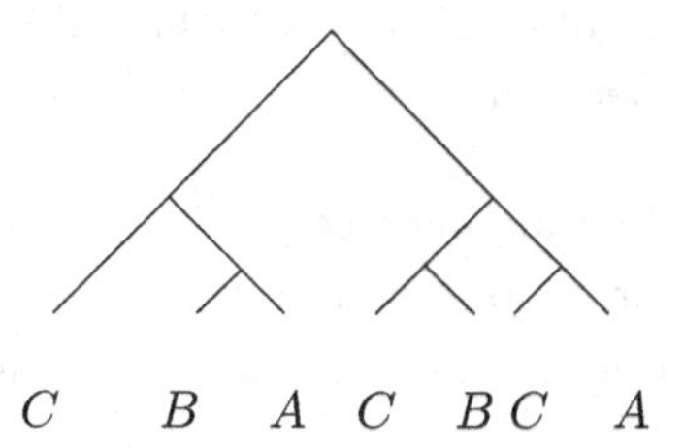

Next annotate the nodes:

$$(C \supset (B \supset A)) \supset ((C \supset B) \supset (C \supset A))$$

$$C \supset (B \supset A) \qquad (C \supset B) \supset (C \supset A)$$

$$B \supset A \quad C \supset B \qquad C \supset A$$

$$C \qquad\qquad B \quad A \; C \quad B \; C \quad A$$

The result is a two-dimensional tree, as in Section 1.5. Therefore:

Frege's logical formulas are syntax trees with missing annotations.

We look next at the rules of proof of Frege's axiomatic logic, in a modern notation and with an anachronistic definition. Proofs begin with axioms, a list of which is given in the end of the *Begriffsschrift*.

Definition 14.3. Logical truth.

(i) *The axioms of logic are logical truths.*
(ii) *If $A \supset B$ and A are logical truths, also B is.*
(iii) *If $A(x)$ is a logical truth for an arbitrary x, also $\forall x A(x)$ is.*

Frege made explicit the principles that govern the notion of an **eigenvariable**, the 'arbitrary' x that is used in mathematics for making universal generalizations. He noticed that a clear-cut syntactic criterion about free variables is sufficient to warrant generalization, which is one of the great insights in the development of logic.

Deduction from the axioms is typically organized in a 'chain of inferences'. In these, any previously derived truths can be used. There is no problem about the combination of derivations, because the premisses of rules are always logical truths. Therefore we have the following features, in contrast to logic after Gentzen:

1. There are no hypothetical inferences.
2. There are no transformations of derivations.
3. The only problem is to find the right instances of axioms.

Finally, let us see how Frege organizes his derivations. The propositional rule of inference is given as:

'*From the two judgments* $\vdash\!\!\begin{array}{l}\top\!\!-A\\ \llcorner\!-B\end{array}$ *and* $\vdash\!-B$ *the new judgment* $\vdash\!-A$ *follows.*'

In the construction of chains of deduction, the writing is linear and one premiss is just referred to by what seems to be a Roman numeral:

$$(X):\ \frac{\vdash\!\!-\!-\ B}{\vdash\!\!-\!-\ A} \qquad\qquad (X)::\ \frac{\vdash\!\!\begin{array}{l}\top\!\!-A\\ \llcorner\!-B\end{array}}{\vdash\!\!-\!-\ A}$$

In these figures, (X): stands for the major premiss $\vdash\!\!\begin{array}{l}\top\!\!-A\\ \llcorner\!-B\end{array}$ and (X):: for the minor premiss $\vdash\!\!-\!-B$.

The **combination** of steps of deduction is shown schematically, as in:

$$(XX,\ XXX)::\ \ \begin{array}{l}\vdash\!\!\begin{array}{l}\top\!\!\top\!\!-A\\ \ \ \llcorner\!-B\\ \llcorner\!\!-\!-\Gamma\end{array}\\ \hline\hline \vdash\!\!-\!-\!-A\end{array}$$

There are two unwritten minor premisses, $\vdash\!\!-\!-\Gamma$ and $\vdash\!\!-\!-\ B$. Frege doesn't tell us how to write a derivation if one premiss is a major one (:) and the other a minor (::); perhaps :,:: would do?

There is only one place in which a combined deduction is used in the whole of the *Begriffsschrift* (viz., in the derivation of formula 102). Nevertheless, even **one place** is proof enough that the thing **exists**.

Frege's linear derivations share the feature with other such deductive systems that premisses have to be referred to by some device, or recalled by a rule of repetition as in Aristotle: In all, we cannot do better than report the words of Roy Dyckhoff who once exclaimed:

Frege had two-dimensional formulas and one-dimensional derivations when he should have had exactly the opposite!

(b) Peano's symbolic notation. Giuseppe Peano's little book of 1889 was written in Latin, with the title *Arithmetices principia, nova methodo exposita*, or 'The principles of arithmetic, presented by a new method'. Peano writes:

> I have denoted by signs all ideas that occur in the principles of arithmetic, so that every proposition is stated only by means of these signs.
>
> ⋮
>
> With these notations, every proposition assumes the form and the precision that equations have in algebra; from the propositions thus written other propositions are deduced, and in fact by procedures that are similar to those used in solving equations.

Peano's signs are, first of all, *dots* that are used in place of parentheses, and then: P for *proposition*, $a \cap b$, even abbreviated to ab, for *the simultaneous affirmation of the propositions a and b*, $-a$ for *negation*, $a \cup b$ for *or*, V for *truth*, and the same inverted for *falsity*. The letter C stands for *consequence*, used inverted as in today's stylized implication sign $a \supset b$. There is also the connective of *equivalence*, $a = b$, definable through implication and conjunction as $a \supset b . \cap . b \supset a$.

Pure logic is followed by a chapter on *classes*, or sets as one could say. The notation is $a \varepsilon b$ for *a is a b*, and $a \varepsilon K$ for *a is a class*.

When Peano proceeds to arithmetic, he first adds to the language the symbols N (*number*), 1 (*unity*), $a + 1$ (*a plus* 1), and $=$ (*is equal to*). The reader is warned that the same symbol is used also for logic. Next he gives the famous Peano axioms for the class N of natural numbers:

Table 14.11 Peano's axioms for natural numbers

1. $1 \varepsilon N$
2. $a \varepsilon N . \supset . a = a$
3. $a, b \varepsilon N . \supset : a = b . = . b = a$
4. $a, b, c, \varepsilon N . \supset \therefore a = b . b = c : \supset . a = c.$
5. $a = b . b \varepsilon N : \supset . a \varepsilon N.$
6. $a \varepsilon N . \supset . a + 1 \varepsilon N.$
7. $a, b \varepsilon N . \supset : a = b . = . a + 1 = b + 1.$
8. $a \varepsilon N . \supset . a + 1 - = 1.$
9. $k \varepsilon K \therefore 1 \varepsilon k \therefore x \varepsilon N . x \varepsilon k : \supset_x . x + 1 \varepsilon k :: \supset . N \supset k.$

The reader would have been helped in axioms 2, 7, and especially in 8 with its negated equality, had separate signs for equality of numbers and equivalence of propositions been used. The last axiom is the principle of induction: Let k be a class that contains 1 and for any x, let it contain $x + 1$ if it contains x. Then it contains the class N. The implication has the eigenvariable x of the inductive step as a subscript.

The list of axioms is followed by a definition:

10. $2 = 1 + 1; 3 = 2 + 1, 4 = 3 + 1$; and so forth.

Now follows a list of theorems, the first one with a detailed proof:

11. $2 \varepsilon N$.

Proof.

$P\,1 . \supset :$	$1 \varepsilon N$	(1)
$1\,[a]\,(P\,6) . \supset :$	$1 \varepsilon N . \supset . 1 + 1 \varepsilon N$	(2)
$(1)\,(2) . \supset :$	$1 + 1 \varepsilon N$	(3)
$P\,10 . \supset :$	$2 = 1 + 1$	(4)
$(4).(3).(2, 1{+}1)\,[a,b]\,(P\,5) : \supset :$	$2 \varepsilon N$	(Theorem).

It will be very useful to inspect this proof in detail. The justifications for each step are written at the head of each line so that they together imply the conclusion of the line. The derivation begins with P 1 (proposition 1 in the list of axioms) in the antecedent, justification part of an implication, and $1 \varepsilon N$ in the consequent as the conclusion. The meaning is that from P 1 follows $1 \varepsilon N$. The second line has similarly that from axiom P 6 with 1 substituted for a follows $1 \varepsilon N . \supset . 1 + 1 \varepsilon N$. The next line tells that from the previous lines (1) and (2) follows $1 + 1 \varepsilon N$. The following line tells that definition 10 gives $2 = 1 + 1$. The last line tells that lines (4) and (3) give, by the substitution of 2 for a and $1 + 1$ for b in axiom P 5, the conclusion $2 \varepsilon N$. The order in which (4) and (3) are listed is $2 = 1 + 1$ and $1 + 1 \varepsilon N$. The instance of axiom P 5 used on the last line in the proof is $2 = 1 + 1 . 1 + 1 \varepsilon N : \supset . 2 \varepsilon N$. Thus, we have quite formally in the justification part the expression:

$$(2 = 1 + 1) . (1 + 1 \varepsilon N) . (2 = 1 + 1 . 1 + 1 \varepsilon N : \supset . 2 \varepsilon N)$$

Line (3) is similar: It has two successive conditions in the justification part, namely $(1 \varepsilon N) . (1 \varepsilon N . \supset . 1 + 1 \varepsilon N)$.

There are altogether two instances of logical inference, both written so that the antecedent of an implication as well as the implication itself is in the justification part, and the consequent of the implication as the conclusion of the line.

Each line of inference in Peano therefore has one of the two forms, with b a substitution instance of axiom a in the first:

$$a \supset b$$
$$a \, . \, a \supset b \, :\supset \, b$$

After the first detailed example, Peano starts to use an abbreviated notation for derivations that makes it rather hard to read them. The first derivation is written 'for the sake of brevity' as:

$$\text{P}\,1 \, . \, 1[a](\text{P}\,6) \, :\supset: \, 1 + 1 \, \varepsilon \, N \, . \, \text{P}\,10\,(2, 1+1)\,[a, b]\,(\text{P}\,5) \, :\supset \, \text{Th.}$$

An even shorter notation is given as an alternative:

$$\text{P}\,1 \, . \, (\text{P}\,6) \, :\supset: \, 1 + 1 \, \varepsilon \, N \, . \, \text{P}\,10\,(\text{P}\,5) \, :\supset \, \text{Th.}$$

It is left for the reader to figure out the meaning of the notation. This expression stands for a formula, in modern notation, of the logical form:

$$A \,\&\, (A \supset B) \supset (B \,\&\, (B \supset C) \supset C)$$

Peano's abbreviation turns a derivation from axioms into a single formula in which the axiom instances together imply the theorem.

The structure of derivations in Peano. *Peano's formal derivations consist of a succession of formulas that are:*

(i) *Implications in which an axiom implies its instance.*
(ii) *Implications in which the conjunction of two previously derived formulas a and $a \supset b$ imply b.*

Peano likened his propositions to the equations of algebra and his deductions to the solving of the equations. Rather startlingly, Jean van Heijenoort who edited the book *From Frege to Gödel* that has the first English translation of Peano's work, instead of figuring out what Peano's notation for derivations means, claims in his introduction that there is 'a grave defect. The formulas are simply listed, not derived; and they could not be derived, because no rules of inference are given . . . he does not have any rule that would play the role of the rule of detachment' (van Heijenoort (1967), p. 84). Had he not seen the forms $a \supset b$ and $a \, . \, a \supset b \, :\supset \, b$ in Peano's derivations, the typographical display of steps of axiom instances and implication eliminations with the conclusion b standing out at right, and the rigorous rule of combining the antecedent of each two-premiss derivation step from previously concluded formulas?

Van Heijenoort's unfortunate assessment, and it becomes much worse if one reads further, has undermined the view of Peano's contribution for a long time, when instead Peano's derivations are constructed purely formally, with a notation as explicit as one can desire, by the application of axiom instances and implication eliminations.

(c) **Russell's synthesis of Frege's logic and Peano's notation.** The pursuit of truth in logic had begun with Gottlob Frege and was continued by Russell and the rest, so that axiomatic logic had become the norm by the 1920s. The axioms were supposed to express the most basic logical truths, and there were just two rules of proof in the passage from instances of axioms to the theorems of logic. The latter were, supposedly, the less basic truths, but sometimes theorems were simpler than axioms. The rules of passage were detachment, from $A \supset B$ and A to conclude B, and the rule of universal generalization. There was another aspect to axiomatic logic, namely a fundamental relativity in the choice of the basic notions. Russell's standard axiomatization of propositional logic, from the *Principia Mathematica*, uses disjunction and negation as the primitive connectives. A slightly modernized notation is:

Table 14.12 *Principia Mathematica* style logical axioms

1. $\neg(A \vee A) \vee A$
2. $\neg A \vee (A \vee B)$
3. $\neg(A \vee B) \vee (B \vee A)$
4. $\neg(A \vee (B \vee C)) \vee ((A \vee B) \vee C)$
5. $\neg(\neg A \vee B) \vee (\neg(C \vee A) \vee (C \vee B))$

Implication is defined by $A \supset B \equiv \neg A \vee B$ and its use would make the above axioms look a little less bad. Say, the first axiom reads as $A \vee A \supset A$ and the last one as $(A \supset B) \supset (C \vee A \supset C \vee B)$. The choice of axioms in Russell and Whitehead is motivated by the algebraic tradition of logic of Ernst Schröder, with such algebraic properties of an operation as idempotence (axiom 1), commutativity (3), and associativity (4).

In a paper of 1906, *The theory of implication*, Russell uses negation and implication as primitives, with a much better-looking axiomatization as a result:

Table 14.13 Russell's 1906 theory of implication

1. $A \supset A$
2. $A \supset (B \supset A)$
3. $(A \supset B) \supset ((B \supset C) \supset (A \supset C))$
4. $(A \supset (B \supset C)) \supset (B \supset (A \supset C))$
5. $\neg\neg A \supset A$
6. $(A \supset \neg A) \supset \neg A$
7. $(A \supset \neg B) \supset (B \supset \neg A)$

The axioms are nicely motivated by intuitive considerations, partly in reference to Frege, partly to Peano, in whose work they appear. Next to the rule of detachment, there is an explicit rule for taking instances of the axioms in formal derivations.

The new axioms did not help in making the provability of theorems of propositional logic any more apparent to Russell than those based on negation and disjunction. He comments wryly ((1906) p. 159):

In the present article, certain propositions concerning implication will be stated as premisses, and it will be shown that they are sufficient for all common forms of inference. It will not be shown that they are all *necessary*, and it is probable that the number of them might be diminished.

Are the axioms sufficient for $(A \supset (A \supset B)) \supset (A \supset B)$? This is hard to tell, but what is worse, there is no logical content in the statement that the axioms are 'sufficient for all common forms of inference'. After the axioms there follow more than twenty pages of formal derivations to bring home the point about sufficiency. The first one is, with the original lower case notation for atomic formulas, and a fractional notation for substitutions in two axioms called Id and Comm, axioms 1 and 4 ((1906) p. 169):

$$\vdash :. \, p \, . \supset : p \supset q \, . \supset . \, q$$

Dem.

$$\vdash . \, \mathrm{Id}\frac{p \supset q}{p} \, . \supset \vdash : p \supset q \, . \supset . \, p \supset q \qquad (1)$$

$$\vdash . \, \mathrm{Comm}\frac{p \supset q, \, p, \, q}{p, \, q, \, r} \, . \supset \vdash :: p \supset q . \supset . p \supset q : \supset :. \, p \supset : p \supset q \, . \supset q \qquad (2)$$

$$\vdash . (1) . \supset : \vdash . (2) . \supset \vdash . \mathrm{Prop.}$$

The structure of derivations in Russell is identical to Peano: The first two lines have an axiom that implies its substitution instance, of the form $a \supset b$, the third line has the form $a \supset ((a \supset b) \supset b)$.

The net effect of the study of axiomatic logic was a scandalously unmanageable logical machinery. Thus, Russell and Whitehead put years into the production of hundreds and hundreds of pages of formal logical proofs within the axiomatic system, but had little idea of its properties as a logical calculus. This aspect becomes obvious through the work of Paul Bernays of 1918. He was studying the propositional calculus of the *Principia Mathematica* and wanted to prove the mutual independence of the axioms, exactly as one does in axiomatic geometry *à la Hilbert.* To this effect, he invented interpretations of the axioms of logic with more than two truth values. Then, certainly unexpectedly, the fourth axiom resisted attempts at proving its independence by such a semantical method. Bernays began suspecting that it could be a theorem and managed to find a derivation for it from the rest of the axioms.

Axiomatic logic had, in the hands of Russell, moved far away from its origin in Frege, namely from a formula language for arithmetic in which, as Frege wrote, 'everything necessary for a correct inference is expressed in full, but what is not necessary is generally not indicated'. Frege, in contrast to Russell, had a clear idea of the practical meaning of his axioms.

14.4 Axiomatic logic in the 1920s

(a) **Hilbert and Bernays: Logical axioms** *à la géométrie.* In the study Bernays made in 1918, what is called a *Habilitationsschrift* or a kind of second doctorate for those who stay at a university, a step back from the relativism of the Russellian axiomatizations had been taken, in that all the standard connectives, conjunction, disjunction, implication, and negation, were present. Bernays hoped to pinpoint the role of negation in logical axiomatics, motivated by the possibility of using negation and any one of conjunction, disjunction, or implication as basic connectives. He asks whether 'it is possible to set up a negation-free axiom system from which all negation-free formulas provable in our calculus, and only these, can be derived'. A negation-free system is one in which conjunction and implication are 'not abbreviations, but symbols for basic operations . . . the question has a positive answer'. Instead of giving the answer, Bernays sets out to derive the fourth of the Russell–Whitehead axioms and to show that the rest are mutually independent.

The formal derivations are written in a linear form, with a stacking of the premisses of the rule of detachment before an inference line. Disjunction is just concatenation:

$$\frac{\begin{cases}\overline{\alpha}\,\beta \\ \alpha\end{cases}}{\beta} \qquad\qquad \frac{\begin{cases}\alpha \to \beta \\ \alpha\end{cases}}{\beta}$$

The notation at right is given as an alternative.

When formal derivations start appearing in Bernays' work, they have always the major premiss of the rule first, then the minor, and the conclusion below the inference line. Repeated applications of the rule are displayed so that this arrangement is kept. It of course happens soon that not all derivations can be arranged under the pattern, unless previously derived formulas are numbered and can be used as premisses in rule instances. A good example in this respect is the derivation of Russell and Whitehead's axiom 4, as in Table 14.12. The very first derivation of the paper is of the formula $X \to YX$, i.e., of $\overline{X}(YX)$ by axioms 2, 3, and 5. The first line in the derivation comes from 'basic formula 5) with XY substituted for X, YX for Y, and $\overline{X}$ for Z':

$$\frac{\dfrac{\begin{cases}(XY \to YX) \to \{\overline{X}(XY) \to \overline{X}(YX)\} \\ XY \to YX \qquad \text{(basic formula 3)}\end{cases}}{\begin{cases}\overline{X}(XY) \to \overline{X}(YX) \\ \overline{X}(XY) \qquad \text{(basic formula 2)}\end{cases}}}{\overline{X}(YX)}$$

A two-premiss rule of syllogism, the use of which is indicated by a double curly bracket, is shown to be derivable:

$$\frac{\left\{\begin{cases}\alpha \to \beta \\ \beta \to \gamma\end{cases}\right.}{\alpha \to \gamma}$$

In the proof of derivability of this rule, the premisses appear as formulas assumed to be derivable, so it is in fact an admissible rule; a notion weaker than derivability. In the rest of the derivations, this rule replaces the rule of detachment.

The negationless axioms Bernays alludes to can be gathered from various sources. One place is the first volume of the *Grundlagen der Mathematik* of 1934 (p. 66).

Table 14.14 The Hilbert–Bernays axioms

1. $A \to (B \to A)$
2. $(A \to (A \to B)) \to (A \to B)$
3. $(A \to B) \to ((B \to C) \to (A \to C))$
4. $A\&B \to A, \quad A\&B \to B$
5. $(A \to B) \to ((A \to C) \to (A \to B\&C))$
6. $A \to A \vee B, \quad B \to A \vee B$
7. $(A \to C) \to ((B \to C) \to (A \vee B \to C))$
8. $(A \to B) \to (\neg B \to \neg A)$
9. $A \to \neg\neg A$
10. $\neg\neg A \to A$

These axioms appear beautiful to anyone with experience in elementary logic. The first three are, in terms of sequent calculus, axiomatic equivalents to the rules of weakening, contraction, and cut. Then follow what look like axiomatic equivalents to the rules of natural deduction for conjunction and disjunction, then the axioms of negation. The implicational axioms go back to Frege.

Bernays wrote in 1918 that there is an axiom system for his negation-free fragment of classical propositional logic. If so, axioms 1–7 of Table 14.14 are not sufficient, for they axiomatize the negationless fragment of intuitionistic logic that falls short of the corresponding classical fragment, witness a purely classical theorem such as $(A \supset B) \vee (B \supset A)$.

(b) Heyting's intuitionistic logic. Discussions about the logic of Brouwer's intuitionistic mathematics were conducted on the pages of the *Bulletin* of the Royal Belgian Academy during the last years of the 1920s. In the midst of these discussions, Heyting figured out in 1928 the proper axiomatization of intuitionistic propositional and predicate logic. There were three predecessors, all unknown to him. In chronological order, Skolem had found in 1919 a structure that is today known as Heyting algebra, one that relates to intuitionistic logic in the same way as Boolean algebra to classical logic, as discussed in Section 14.2(c). The matter was known to Tarski and others when they, in the latter part of the 1930s, figured out the algebraic semantics of intuitionistic logic (see von Plato (2007) for details). A second precursor to Heyting was Kolmogorov in his Russian paper of 1925. A third precursor was Bernays, who after a talk by Brouwer in Göttingen found out in 1925 that it is sufficient to leave out the law of double negation from a suitable axiomatization of classical logic, like the one in Table 14.14.

Brouwer was enthusiastic towards Heyting's work, and concluded that he would not have to finish such a work himself. There has been for some reason a general belief that Brouwer was somehow against the formalization of intuitionistic logic, but this is a clearly erroneous idea.

Heyting's axiomatization has eleven propositional axioms, not quite like those of *Principia Mathematica* as has been often said, but more like the Hilbert–Bernays axioms in which all connectives and quantifiers are present. The original version of 1928 is lost, but the published version refers to a paper by V. Glivenko of 1928 that has a handsome set of axioms.

Implication elimination was not the only propositional rule, but there was also an explicit rule of conjunction introduction. It could be dispensed with, but the proofs would then be 'even more intricate' (*verwickelt*). A little reconstruction shows that this is the case:

Heyting's first theorem is $\vdash \cdot a \wedge b \supset a$, and the proof he gives is:

[2.14] $\vdash \cdot a \supset \cdot b \supset a :\supset:$
[2.12] $\vdash \cdot a \wedge b \supset \cdot b \supset a \cdot \wedge b \cdot$ [2.15] b [should be a]

The notation uses abbreviations and the dot notation in place of parentheses, with the axioms referred to by numbers in square brackets. When axioms are written, they are indicated by a double turnstile, and a more detailed rendering of Heyting's derivation, by his own conventions, would be:

[2.14] $\vdash\vdash \cdot a \supset \cdot b \supset a$
[2.12] $\vdash\vdash \cdot a \supset \cdot b \supset a :\supset: a \wedge b \supset \cdot b \supset a \cdot \wedge b$
$\vdash \cdot a \wedge b \supset \cdot b \supset a \cdot \wedge b$
[2.15] $\vdash\vdash \cdot b \wedge \cdot\ b \supset a \cdot \supset a$

The axioms in use here are:

[2.14] $\vdash\vdash \cdot b \supset \cdot a \supset b$
[2.12] $\vdash\vdash \cdot a \supset b \cdot \supset \cdot a \wedge c \supset b \wedge c$
[2.15] $\vdash\vdash \cdot a \wedge \cdot\ a \supset b \cdot \supset b$

A complete proof without abbreviations has eleven lines, the formulas written for clarity with parentheses. I have organized it so that each rule instance is preceded by its premisses in a determinate order. [1.3] is implication elimination, [1.2] conjunction introduction, the rest are axiom instances:

1. [2.14] $\vdash a \supset (b \supset a)$
2. [2.12] $\vdash (a \supset (b \supset a)) \supset (a \wedge b \supset (b \supset a) \wedge b))$
3. [1.3] $\vdash a \wedge b \supset (b \supset a) \wedge b$
4. [2.11] $\vdash (b \supset a) \wedge b \supset b \wedge (b \supset a)$

5. [1.2] $\vdash (a \wedge b \supset (b \supset a) \wedge b) \wedge ((b \supset a) \wedge b \supset b \wedge (b \supset a))$
6. [2.13] $\vdash ((a \wedge b \supset (b \supset a) \wedge b) \wedge ((b \supset a) \wedge b \supset b \wedge (b \supset a)) \supset (a \wedge b \supset b \wedge (b \supset a))$
7. [1.3] $\vdash a \wedge b \supset b \wedge (b \supset a)$
8. [2.15] $\vdash b \wedge (b \supset a) \supset a$
9. [1.2] $\vdash (a \wedge b \supset b \wedge (b \supset a)) \wedge (b \wedge (b \supset a) \supset a)$
10. [2.13] $\vdash ((a \wedge b \supset b \wedge (b \supset a)) \wedge (b \wedge (b \supset a) \supset a)) \supset (a \wedge b \supset a)$
11. [1.3] $\vdash a \wedge b \supset a$

A reader who undertakes to turn Heyting's proof outlines into formal derivations will soon notice two things:

1. There is so much work that one lets it be after a while.

2. The way to proceed instead becomes clear, because one starts to read Heyting's axioms and theorems in terms of their intuitive content.

Formally, axioms or previously proved theorems of the forms $a \supset b$ and a lead to another theorem b by a step of implication elimination. In practice, one sees implications $a \supset b$ as rules by which b can be concluded whenever a is at hand, as Heyting seems to have done, considering his very sketchy derivations. The situation is particularly tempting if the major premiss is an axiom such as $a \supset a \vee b$. The step to deleting the axiom and implication elimination, to conclude $a \vee b$ directly from a, is short, as in:

1. $\vdash\vdash a \supset a \vee b$
2. $\vdash a$
3. [1.3] $\vdash a \vee b$

It would suffice to delete line 1 (as Frege's notation does) and to change the reference to rule [1.3] into an explicit disjunction introduction rule. Thus, with the experience of actually using Heyting's logical calculus for the construction of formal proofs, one reads his first theorem $a \,\&\, b \supset a$ as: From $a \,\&\, b$ follows a. Axioms such as $a \supset a \vee b$, $b \supset a \vee b$, and $(a \supset c) \supset ((b \supset c) \supset (a \vee b \supset c))$ turn into what are now the familiar natural deduction rules of disjunction. With these observations, we have arrived in our survey of deductive systems at the precise point at which Gentzen began his development of natural deduction.

Suggestions for the use of this book

Basic design: This book is based on lecture courses I have held during many years at the University of Helsinki. Most students have come from the humanities and experienced their one and only contact with the modes of thinking of the exact sciences in this first logic course. Others came from the natural sciences and often had an easier time, but practically all students have learned to master the basic proof systems of propositional and predicate logic, with the capacity to do almost any standard proofs by themselves. Formal work has been complemented by epistemological discussions of what comes first in conceptual order in logic, and in philosophy more generally, namely, truth or proof, and by occasional glimpses into the history of logic.

There is very little in the beginning by the way of connections between logical and natural language. The teaching moves, instead, at once into logical arguments in basic propositional logic that the students learn to read and understand through a couple of examples. This design feature has been surprisingly unproblematic.

The idea with the logical systems is to start with linear derivations for conjunction, implication, and negation, because it is so easy for the student. Next these derivations are translated into a tree form that shows, contrary to the linear one, what depends on what. The purely algorithmic nature of the translation is emphasized.

The trees of standard natural deduction grow in the wrong way, down from their leaves. Proof strategies are needed that are only partly supported by the notation. A further step to a sequent notation is made, with a final form of the logical calculus in which the construction of formal derivations in the root-first way is enjoyable. Experience has shown that the sequent form would not be a good starting point, because the content of the logical rules would remain unclear. That form is too far removed from intuitive logical reasoning and can, in my experience, be taken into use only gradually.

The view of logic here is that it codes modes of demonstrative argument. The presentation of classical logic and especially its 'truth table' semantics is postponed as far as possible. In my experience, the early introduction of truth tables impairs seriously the learning of logic. My presentation of classical truth value semantics builds on the wonderful fully invertible classical sequent calculus for propositional logic, and one gets a direct touch of the connection between syntactic derivability and logical truth in Section 7.4.

The emphasis in the formal work is on the side of the intuitionistic and classical propositional calculi. My experience is that when they are mastered, predicate logic will be easily learnt. The basic ideas about quantification are presented in detail, especially the meaning of universal quantification through Frege's idea of eigenvariables. The semantics of predicate logic, especially completeness, which traditionally is beyond a first course, is discussed only briefly.

Chapter 11 discusses equality relations, predicate logic with equality as an example, and axiomatic theories with lattice theory as an example. Chapter 12 is dedicated to elementary arithmetic. It is seen how the level of difficulty rises steeply with the last one, even if the presentation is thoroughly elementary.

Chapter 13 complements the presentation with topics such as structural induction and normalization for natural deduction. The Curry–Howard correspondence is briefly presented, to give an idea of the connections between elementary logic and functional programming.

The rule of cut of sequent calculus appears only as a complementary topic. It was somewhat of a surprise that the admissibility of a rule of contraction could be proved in an absolutely elementary way with termination of proof search as a result, once cut was not a concern. There is no need of the property of preservation of derivation height in the elimination of contraction and a reader with good capacities of concentration can go through the proof of Section 4.5.

A final Chapter 14 presents a historical overview of the great deductive machinery works, beginning with the very first one, namely Aristotle's syllogistic. The tour ends with the high point of axiomatic logic, Heyting's system of intuitionistic predicate logic in 1930, right before the miraculous emergence of Gödel's incompleteness results that changed the course of logic. Gentzen's proof systems, on which this presentation of the topic is based, were by-products in his attempt at understanding the meaning of incompleteness. This story is told in my recent 'Gentzen's proof systems: Byproducts in a work of genius' (von Plato 2012).

Exercises are collected at the end of each chapter. Parts of them complement the main text in various ways.

A minimum with options: I have usually had about fifty hours available, including exercises, so not all of the materials could be presented in the lectures. The latter parts of many chapters can be left as reading material.

All of Chapters 11 and 12, except for the mentioning of predicate calculus with equality, is optional material in respect of the central aims of the course.

The book could equally well be used for a leisurely second course with emphasis on proof systems and all of it covered.

Further reading

Here are some suggestions for further reading. The book *Structural Proof Theory* (2001) by Sara Negri and Jan von Plato, available also in paperback since 2008, presents in its first half in detail the contraction-free sequent calculi of Chapters 4, 6, and 9. The second half develops further variants of sequent calculi, natural deduction, and the extension of sequent calculi by rules that correspond to mathematical axioms. This last-mentioned topic is the subject of another book by the same authors titled *Proof Analysis: A Contribution to Hilbert's Last Problem* (2011).

Troelstra and Schwichtenberg's *Basic Proof Theory* (2000) covers a lot of ground but gives its rewards only to those who are willing to put a lot of work into its reading.

Jean van Heijenoort edited a collection of papers on the development of logic, with the title *From Frege to Gödel: A Source Book in Mathematical Logic, 1879–1931* (1967). As its name indicates, it starts with Frege's *Begriffsschrift* and ends up with Gödel's incompleteness paper. A fresh collection of studies is *The Development of Modern Logic* (ed. Haaparanta (2009)), therein especially the concise thematic presentation of the development of mathematical logic by Mancosu, Zach, and Badesa.

Gödel's papers, some manuscripts, and correspondence are found in the five-volume *Collected Works of Kurt Gödel.* His life is detailed in John Dawson's *Logical Dilemmas: The Life and Work of Kurt Gödel* (1997). Gödel's incompleteness theorem is the topic of Peter Smith's very readable *An Introduction to Gödel's Theorems* (2007). An enjoyable and accessible companion to Gödel's work is Torkel Franzén's *Gödel's Theorem: An Incomplete Guide to its Use and Abuse* (2005).

Gentzen's work is available in English translation in his *Collected Papers* (ed. Szabo (1969)). Recently discovered details of his achievement can be found chronicled in von Plato (2009, 2012). His tragic life story is told in Menzler-Trott's *Logic's Lost Genius: The Life of Gerhard Gentzen* (2007).

Enjoyable classics include Kleene's *Introduction to Metamathematics* (1952) and Prawitz' *Natural Deduction: A Proof-Theoretical Study* (1965) (available as a Dover paperback since 2006). Kleene's *Mathematical Logic* (1967) is now similarly available.

The perhaps worst lacuna in logical literature is the lack of an accessible introduction to the proof theory of arithmetic. The best one can do is to try to work oneself through the first half of Takeuti's *Proof Theory* (1975). The recent book *Proof Theory: The First Step into Impredicativity* by Wolfram Pohlers (2009) could then perhaps be tackled by the absolutely motivated readers.

Bibliography

Aristotle, *Prior Analytics*. Several translations.

Bernays, P. (1945) Review of Ketonen (1944). *The Journal of Symbolic Logic*, vol. 10, pp. 127–30.

Beth, E. (1955) Semantic entailment and formal derivability. *Mededelingen der Koninklijke Nederlandse Akademie van Wetenschappen*, Afd. Letterkunde, vol. 18, pp. 309–42.

Boole, G. (1847) *The Mathematical Analysis of Logic, Being an Essay Towards a Calculus of Deductive Reasoning* (repr. Oxford: Blackwell, 1965).

Church, A. (1932) A set of postulates for the foundation of logic. *Annals of Mathematics*, vol. 33, pp. 346–66.

(1936) A note on the Entscheidungsproblem. *The Journal of Symbolic Logic*, vol. 1, pp. 101–2.

Curry, H. and R. Feys (1958) *Combinatory Logic*, vol. 1. Amsterdam: North-Holland.

Dawson, J. (1997) *Logical Dilemmas: The Life and Work of Kurt Gödel*. Oxford University Press.

Dragalin, A. (1978) *Mathematical Intuitionism: Introduction to Proof Theory*. Providence, RI: American Mathematical Society.

Franzén, T. (2005) *Gödel's Theorem: An Incomplete Guide to its Use and Abuse*. Wellesley, MA: A. K. Peters.

Frege, G. (1879) *Begriffsschrift: Eine nach der arithmetischen nachgebildete Formelsprache des reinen Denkens*. English translation in van Heijenoort (1967).

Gentzen, G. (1933) Über das Verhältnis zwischen intuitionistischer und klassischer Arithmetik. Submitted 15 March 1933 but withdrawn, published in *Archiv für mathematische Logik*, vol. 16 (1974), pp. 119–32.

(1934–5) Untersuchungen über das logische Schliessen. *Mathematische Zeitschrift*, vol. 39, pp. 176–210 and 405–31.

(1969) *The Collected Papers of Gerhard Gentzen*, ed. M. Szabo. Amsterdam: North-Holland.

(2008) The normalization of derivations. *The Bulletin of Symbolic Logic*, vol. 14, pp. 245–57.

Gödel, K. (1931) Über formal unentscheidbare Sätze der Principia Mathematica und verwandter Systeme I. *Monatshefte für Mathematik und Physik*, vol. 38, pp. 173–98. Also in Gödel (1986).

(1933a) Zur intuitionistischen Arithmetik und Zahlentheorie. As reprinted in Gödel (1986), pp. 286–95.

(1933b) The present situation in the foundations of mathematics. In Gödel (1995), pp. 45–53.

(1986, 1990, 1995, 2003) *Collected Works*, vols. 1–5. Oxford University Press.

Haaparanta, L., ed. (2009) *The Development of Modern Logic*. Oxford University Press.

Hakli, R. and S. Negri (2012) Does the deduction theorem fail for modal logic? *Synthese*, vol. 187, pp. 849–67.

van Heijenoort, J., ed. (1967) *From Frege to Gödel, A Source Book in Mathematical Logic, 1879–1931*. Cambridge, MA: Harvard University Press.

Heyting, A. (1930) Die formalen Regeln der intuitionistischen Logik. *Sitzungsberichte der Preussischen Akademie von Wissenschaften, Physikalisch-mathematische Klasse*, pp. 42–56.

Hilbert, D. and P. Bernays (1934) *Grundlagen der Mathematik I*. Berlin: Springer.

Howard, W. (1969) The formulae-as-types notion of construction. Published in 1980 in J. Seldin and J. Hindley, eds., *To H. B. Curry: Essays on Combinatory Logic, Lambda Calculus and Formalism*, pp. 480–90. London and New York: Academic Press.

Ketonen, O. (1944) *Untersuchungen zum Prädikatenkalkül* (Annales Academiae Scientiarum Fennicae, Ser. A.I. 23).

Kleene, S. (1952), *Introduction to Metamathematics*. Amsterdam: North-Holland.

(1967) *Mathematical Logic*. New York: Dover.

Kolmogorov, A. (1932) Zur Deutung der intuitionistischen Logik. *Mathematische Zeitschrift*, vol. 35, pp. 58–65. English translation in Mancosu (1997).

Kripke, S. (1965) Semantical analysis of intuitionistic logic I. In J. Crossley and M. Dummett, eds., *Formal Systems and Recursive Functions*, pp. 92–130. Amsterdam: North-Holland.

Mancosu, P. (1997) *From Brouwer To Hilbert: The Debate on the Foundations of Mathematics in the 1920s*. Oxford University Press.

Mancosu, P., R. Zach, and C. Badesa (2009) The development of mathematical logic from Russell to Tarski, 1900–1935. In Haaparanta (2009), pp. 318–470.

Martin-Löf, P. (1984) *Intuitionistic Type Theory*. Naples: Bibliopolis.

Menzler-Trott, E. (2007) *Logic's Lost Genius: The Life of Gerhard Gentzen*. Providence, RI: American Mathematical Society.

Negri, S. and J. von Plato (2001) *Structural Proof Theory*. Cambridge University Press.

(2011) *Proof Analysis: A Contribution to Hilbert's Last Problem*. Cambridge University Press.

Peano, G. (1889) *Arithmetices principia, nova methodo exposita*. English translation in van Heijenoort (1967).

von Plato, J. (2001) Natural deduction with general elimination rules. *Archive for Mathematical Logic*, vol. 40, pp. 541–67.

(2007) In the shadows of the Löwenheim–Skolem theorem: Early combinatorial analyses of mathematical proofs. *The Bulletin of Symbolic Logic*, vol. 13, pp. 189–225.

(2009) Gentzen's logic. In D. Gabbay and J. Woods, eds., *Handbook of the History of Logic*, vol. 5, *Logic from Russell to Church*, pp. 667–721. Amsterdam: Elsevier.

(2012) Gentzen's proof systems: Byproducts in a work of genius. *The Bulletin of Symbolic Logic*, vol. 18, pp. 313–67.

(2014) From *Hauptsatz* to *Hilfssatz*. In M. Baaz, R. Kahle, and M. Rathjen, eds., *The Quest of Consistency* (in press).

von Plato, J. and A. Siders (2012) Normal derivability in classical natural deduction. *The Review of Symbolic Logic*, vol. 5, pp. 205–11.

Pohlers, W. (2009) *Proof Theory: The First Step into Impredicativity*. Berlin: Springer.

Prawitz, D. (1965) *Natural Deduction: A Proof-Theoretical Study*. Stockholm: Almqvist & Wicksell (repr. New York: Dover, 2006).

(1991) *ABC i Symbolisk Logik* (in Swedish), 2nd edn. Stockholm: Bokförlaget Thales.

Prior, A. (1969) The runabout inference-ticket. *Analysis*, vol. 21, pp. 38–9.

Raggio, A. (1965) Gentzen's Hauptsatz for the systems NI and NK. *Logique et analyse*, vol. 8, pp. 91–100.

Russell, B. (1906) The theory of implication. *American Journal of Mathematics*, vol. 28, pp. 159–202.

Siders, A. (2014) A direct Gentzen-style consistency proof for Heyting arithmetic. In M. Baaz, R. Kahle, and M. Rathjen, eds., *The Quest of Consistency* (in press).

Skolem, T. (1920) Logisch-kombinatorische Untersuchungen über die Erfüllbarkeit oder Beweisbarkeit mathematischer Sätze nebst einem Theoreme über dichte Mengen. As reprinted in Skolem (1970), pp. 103–36.

(1923) Einige Bemerkungen zur axiomatischen Begründung der Mengenlehre. As reprinted in Skolem (1970), pp. 137–52. English translation in Van Heijenoort (1967).

(1929) Über einige Grundlagenfragen der Mathematik. As reprinted in Skolem (1970), pp. 227–73.

(1970) *Selected Works in Logic*, J. E. Fenstad. Oslo: Universitetsforlaget.

Smith, P. (2007) *An Introduction to Gödel's Theorems*. Cambridge University Press.

Takeuti, G. (1975) *Proof Theory*. Amsterdam: North-Holland.

Troelstra, A. and H. Schwichtenberg (2000) *Basic Proof Theory*, 2nd edn. Cambridge University Press.

Turing, A. (1936) On computable numbers, with an application to the Entscheidungsproblem. *Proceedings of the London Mathematical Society*, vol. 42, pp. 230–65.

Whitehead, A. and B. Russell (1910–13) *Principia Mathematica*, vols. I–III. Cambridge University Press.

Index of names

Index of subjects

Printed in the United States
By Bookmasters

Printed in the United States
By Bookmasters